THE SECRET LIFE OF STARS

ASTROPHYSICS FOR EVERYONE

THE SECRET LIFE OF STARS

LISA HARVEY-SMITH

To the children, whose eyes are wide open.
And to the adults who refuse to grow up.

First published in Australia in 2020
by Thames & Hudson Australia Pty Ltd
11 Central Boulevard, Portside Business Park
Port Melbourne, Victoria 3207
ABN: 72 004 751 964

First published in the United States of America in 2021. Reprinted in 2021.
by Thames & Hudson Inc.
500 Fifth Avenue
New York, New York 10110

The Secret Life of Stars © Thames & Hudson Australia 2020

Text © Lisa Harvey-Smith 2020
Illustrations © Eirian Chapman 2020

23 22 21   5 4 3

Thames & Hudson Australia wishes to acknowledge that Aboriginal and Torres Strait Islander people are the first storytellers of this nation and the traditional custodians of the land on which we live and work. We acknowledge their continuing culture and pay respect to Elders past, present and future.

ISBN 978-1-760-76122-6 (AUS hardback)
ISBN 978-1-760-76158-5 (U.S. hardback)
ISBN 978-1-760-76136-3 (ebook)

A catalogue record for this
book is available from the
National Library of Australia

Library of Congress Control Number 2020939572

Every effort has been made to trace accurate ownership of copyrighted text and visual materials used in this book. Errors or omissions will be corrected in subsequent editions, provided notification is sent to the publisher.

Cover Illustration: Eirian Chapman
Cover Design: Philip Campbell Design
Typesetting: Cannon Typesetting
Editing: Katie Purvis
Printed and bound in China by C&C Offset Printing Co., Ltd.

Be the first to know about our new releases,
exclusive content and author events by visiting
thamesandhudson.com.au
thamesandhudsonusa.com
thamesandhudson.com

FSC® is dedicated to the promotion of responsible forest management worldwide. This book is made of material from FSC®-certified forests and other controlled sources.

Contents

Starry starry night

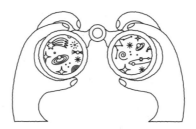

There are more than 7 billion human beings on planet Earth. We are blessed with a large variety of skin tones, eye colours and hair colours, not to mention a fascinating diversity of earlobe shapes (if you don't believe me, do an internet search).

We come in many shapes and sizes and have different abilities, interests and personalities. Some characteristics are encoded in our DNA from birth, but many aspects of our personalities and physical make-up change profoundly as we age.

Some of us find solace in the arts, music, poetry and language; others' minds are fed and fascinated by the sciences. My chosen field of astrophysics is a discipline rooted in both art and science, for the beauty and unpredictability of the cosmos stimulates every part of our brains. After all, the stars in our night sky make up the greatest and most detailed artistic canvas in the universe.

There are probably around a billion trillion stars in the observable part of the universe – that's 1,000,000,000,000,000,000,000 for you visual types. We have no idea how many there might be in the rest of our cosmos because the expansion of it means that the light from those stars will never reach the Earth.

Of those billion trillion stars, only about 3000 are visible to the naked eye. To our gaze, whether fleeting or devoted, it seems there is little to separate them. From the brilliant white Sirius to the intense orange stare of Aldebaran to the countless anonymous faint ones whose names we don't know, there is not much (apart from their brightness and colour) to separate each tiny point of light.

Luckily, humans are smart, and we have developed tools to magnify, deconstruct and analyse the light that comes from the heavens. An avalanche of new information is continually gathered, from their temperature to their chemical make-up and their age. The past behaviours and tantrums of stars (think explosions and eruptions) are scrutinised from records of long-dead scientists, and their likely future actions are predicted by the patterns of deeds exhibited by other stars.

Astrophysics is a detailed science. When we adopt such a scientific approach, many of our stellar subjects that seem normal and predictable when we glance up on a clear night reveal their true nature. The fact is, our universe is home to a whole host of temperamental personalities. We see stable dwarf stars, unpredictable giants, and many in between. We see kind stars, devious stars, selfish and just plain weird stars.

Some live in families, yet many destroy their relationships or even kill and eat their partners. During a midlife crisis a star can disappear completely, or reincarnate in a colourful cloud of gas. Stars are born and they age, just like us, before slowly succumbing to the inevitable, their ashes returned to the cosmos.

As we travel the universe using the vehicle of science, we discover incredible new things we could never have imagined. We peer through their keyholes to see how they live when nobody's looking. Nothing is sacred, except the laws of physics – and even they can sometimes be negotiable.

Welcome to the secret lives of stars.

So erm, what exactly is the Sun?

Unless you're less than six months old and live at the South Pole, I'm guessing you're familiar with the Sun, our nearest star. At around 4.6 billion years old, the Sun is in the middle age of her life. And before you ask, yes, the Sun is a woman. How do I know? She holds down a steady job (heating and lighting the solar system), provides for a family of eight and hasn't ever taken a holiday.

Our star is made up of the lightest gas in the universe, called hydrogen, as well as bits and bobs of other chemicals that she collected from outer space when she was moulded into a ball by the force of gravity.

Hydrogen is the most common element in the universe. If you zoom in millions of times, 'Honey-I-shrunk-the-kids'-style, each atom of hydrogen is made from one tiny particle called a proton and one even tinier particle called an electron. Protons have a positive electrical charge and electrons have a negative charge, and opposites attract – so the force between the proton and electron keeps the atom together.

Inside the Sun, things are a little different. The solar gas is so hot and the atoms are shaken so ferociously that they are broken into pieces, with protons and electrons left whirling about as if in a savage dust storm. Where all the atoms are broken like this, the resulting gas is called a plasma. So really, the environment of the Sun is like the inside of one of those fun zappy plasma balls where you press the glass sphere with your finger and a tiny bolt of lightning touches it, discharging a purple arc of light. But never touch the Sun, y'all – it's hot.

We depend on the Sun more than we can ever imagine. As the source of all heat and light on Earth, our star is completely essential to life on our planet. She doesn't burn like a fire but shines with her own power – a power that has barely diminished in thousands of millions of years. Her life force comes from deep within her, as if she were a radiant Buddha. Instead of karmic energy, though, what she releases is more like a violent explosion of heat, light and dangerous radiation that, if you got too close, could burn you to a crisp faster than you can say 'Pass the factor 50'.

Why does the Sun shine? Well, it's a real melee inside her, with gas crushed together causing a rumbling and grumbling that is symptomatic of something a little stronger than heartburn. In the middle of our star, the gas is all squished together by the massive weight of material above it. It's a bit like the pressure we feel due to the weight of the atmosphere. At sea level, we have around 100 kilometres of atmosphere on our heads. As a result of this enormous mass bearing down on us, every square centimetre of our body feels a force equivalent to 1 kilogram from the weight of the air. At the top of Mount Everest, where most of the atmosphere is below your feet, this weight of air drops to less than 300 grams.

Since the Sun is made of gas, or, strictly, plasma, it's basically one big atmosphere all the way through. Don't think clear blue skies, though. The radius of the Sun is around 700,000 kilometres from the core to the surface, and the weight of all this gas causes an enormous pressure in the core of about 3.8 trillion PSI (a million million times your car's tyre pressure). In the Sun's core, the so-called 'gas' has around 150 times the density of water, an amount that is denser than lead!

Early on in the Sun's life, this colossal pressure squashed and heated up the plasma to such a level that it triggered a spectacular subatomic cooking show that has continued unabated ever since. The temperature in the Sun's core is 15 million degrees Celsius – so it's not a great place for a holiday. This unimaginable heat is generated by the nuclear reactions that produce all the Sun's energy and light. In the raging fire of this dense and hot furnace, tiny particles crash into each other and stick together. Quickly, pieces of atoms are formed along with lots of heat and light. It's like an atomic jigsaw puzzle.

Interested in the gory details? Here goes. The Sun shines by a process called nuclear fusion, which happens when a bunch of protons and neutrons crash into each other and decide to make a complicated, messy group.

We know that the proton is the building block at the centre of a hydrogen atom; it's also called hydrogen's atomic nucleus. In a series of collisions, six protons combine to make a pair of protons and a helium nucleus made from two protons and two neutrons. So that's six protons in, and four protons and two neutrons out. This reaction is happening 100 trillion trillion trillion (100,000,000,000,000,000,000,000,000,000,000,000) times every second in the Sun and in most of the 1 billion trillion (1,000,000,000,000,000,000,000,000) other stars in our universe.

How does it work? Where did the other two protons go, and where did the pair of neutrons come from?

It turns out that particles are pretty flexible creatures. They can be converted into other types, or even into energy, so long as a bunch of rules are followed. For every nuclear reaction that happens in the Sun, a shower of smaller particles is created too. They include subatomic weirdos like neutrinos (particles thought to have no mass) and positrons (particles of antimatter – yikes!)

and cosmic rays. Many of these particles are coursing through your body right now, harmlessly emerging out the other side of your skin, but some of the more dangerous could potentially cause damage to your DNA. By a stroke of luck, these baddies are mostly deflected by the magnetic force field of the Earth and never enter our 'hood.

The other important thing produced by nuclear fusion in the Sun is LOADS of energy, in the form of gamma rays. That's where sunshine really comes from.

Gamma rays are the most energetic (and, to humans, deadly) type of radiation that exists on the spectrum of light and colour. The nuclear reactions in the Sun give out only gamma rays, and no light at all. The more friendly and familiar forms of radiation – heat and light – that we see from the Sun come later, when the gamma rays bump into particles inside the Sun, are absorbed and are re-emitted in a random direction with a slightly lower energy as the collision 'steals' some of the energy of each incoming ray. The process is so haphazard that it takes a gamma ray emitted by nuclear fusion around 30,000 years to get from the core of the Sun to the surface as heat and light. We call the process, amusingly, 'random walk'.

That, my friend, is how the Sun shines.

It's amazing to realise how vast the quantities of energy being produced from the gas that makes up our Sun are. A mind-bending 4 million tonnes of hydrogen plasma is converted into energy every second inside the Sun. That's enough to service the current power consumption of the Earth for more than 4 trillion years.

You might think that nuclear fusion would be a great way to generate our power requirements here on Earth, and in some ways you'd be right. Nuclear fusion would have none of the dangerous by-products of nuclear fission, which is used in the 'splitting the atom' technology currently employed in nuclear power stations. Unfortunately, though, we haven't yet mastered the extreme conditions required to initiate and sustain nuclear fusion reactions safely here on Earth.

The processes that generate energy in the Sun are the same as the ones going on inside a thermonuclear bomb. Perhaps you're wondering: 'If the Sun is a giant thermonuclear bomb, how come it doesn't just explode?'

The answer is simply 'Gravity keeps it in'.

Yep – believe it or not, the humble force of gravity keeps the phenomenal 'H-bomb' at the core of the Sun from exploding into space. The outward forces of searing-hot, ever-expanding gases are balanced perfectly by the sheer weight of the star. Our Sun is as one, balanced and centred like a yoga goddess.

That's not to say she is completely static, though. Her outer layer, called the convection zone, is a thick layer of gas at more than 1 million degrees Celsius that rises to the surface, then cools to a mere 6000 degrees and falls back down below like globs of boiling soup on a hot stove.

This bubbling cauldron of heated gas beneath the surface of the Sun affects how she spins. A 'day' on the Sun (in other words, how fast she spins once on her head) ranges from 25 to 36 Earth days, depending on whether you are at her equator or her poles. This is quite unlike the spin of the Earth, which takes exactly 23 hours, 56 minutes and 4 seconds wherever you happen to be standing. The hot fluid convection cells rising from beneath the surface of the Sun promote a quicker spin at the equator and a slower rate of rotation at her poles. This weird 'differential' rotation causes her magnetic field, which is generated deep within her belly, to tangle up as the faster bits overtake the slower bits. The Sun, in other words, has knots in her stomach.

Where does this magnetic field come from?

We know it's been there for a very long time, perhaps even before the Sun started shining. We also know that her magnetic field has remained strong for billions of years. We have good reason to believe that it is generated by something called a dynamo, deep within the Sun.

Ever heard of a dynamo? If you're as old as me, you might remember dynamo lights on bicycles – environmentally friendly, with no batteries needed. (Perhaps we should go back to using them.)

Dynamo lights work by placing a small wheel against one of the bike's tyres. As the wheel turns, a magnet inside it spins around. A coil of copper wire is placed next to the spinning magnet and, as its force field rotates, a small electrical current is created within the wire, which powers the bicycle light. Electricity and magnets

THE SECRET LIFE OF STARS

arc connected: when one moves, the other is induced. That's what we call 'electromagnetism'.

A similar effect (but in the opposite direction) is happening inside the Sun. Magnetic forces are created inside our star from the action of spinning electrical currents.

I know what you're thinking. Why are there electrical currents inside the Sun in the first place? Has someone wired her up and plugged her in – some sort of intrepid sparkie with asbestos hands?

Well, no, but the movement of hot gases in the Sun, which spin with the rotation and convection just under the surface, actually amounts to an electrical current. Electricity is just the motion of charged particles. The gas within the Sun is so hot that the atoms are broken up into their constituent pieces, many of which (for example, protons and electrons) are electrically charged. All this positive and negative charge swirling around generates a magnetic field that is about twice as strong as our Earth's.

Because different parts of the Sun rotate at different speeds, the magnetic field is squeezed and tangled up over time. In some places, it becomes 8000 times stronger than the Earth's puny magnetic field. This causes some rather drastic heliospheric acne, called sunspots. They look like black dots on the surface of our star. That's because the hot gases that usually bubble up from below are pushed under by the tremendous magnetic forces in the sunspot. These parts of the Sun are therefore much cooler, around 4600 degrees, and so appear darker than the rest of the burning orb.

Another symptom of the Sun's magnetic field is solar flares. These gigantic eruptions, or burps, from our star originate when the tangled magnetic field of a large group of sunspots suddenly 'snaps' and rights itself, releasing vast quantities of hot plasma from below the surface. It's a bit like when a rubber band gets twisted and knotted, then finally breaks.

Solar flares can be seen with special filtered telescopes as soon as they happen, but the effects on Earth are not felt until several days later. These solar eruptions – with 10 million times the energy of a volcanic eruption – spew searing-hot plasma into space. The gas escaping can reach temperatures of up to 100 million degrees.

The most powerful solar flare ever recorded happened on 1–2 September 1859. Vast globs of solar plasma rocketed towards the Earth at unprecedented speeds, taking only seventeen-and-a-half hours to reach the planet. The skies were alight with glowing green and red light as the electrically charged stuff spilled into our atmosphere and interacted with the oxygen and nitrogen in the air, causing one of the brightest auroral displays ever recorded.

Usually, the aurora would be confined to the north and south polar regions because solar plasma is guided that way by the Earth's magnetic field. This event, however, was so powerful that the solar plasma spilled down to tropical regions, lighting up the skies above Hawaii, Queensland, the Caribbean and Colombia. There were reports in a US gold mine of workers getting up and preparing breakfast in the middle of the night since the sky was so bright that they believed it was already morning.

With the storm came disturbances to the (relatively new) telegraph systems, which used electrical signals in wires to communicate, much as we do today with our landline telephones. As the electrically charged gas from the Sun streamed through the atmosphere, telegraph systems were overwhelmed by the additional electrical currents. Telegraph operators received electric shocks, systems went down intermittently over a period of two days, and pieces of paper placed adjacent to telegraph equipment reportedly caught fire. In places, the currents were so high that some messages were sent even after the electricity was cut to the systems!

There have been other major solar storms since, but none so large as that of 1859. In 1989, an eruption knocked out the electricity grid of Quebec for more than nine hours, overwhelming the province's powerlines by inducing massive electrical currents in the wires. Many satellites, including weather and communications systems, were disrupted, and anomalies were detected on the space shuttle *Atlantis*. Radio stations were jammed and United Nations peacekeeping operations in Namibia were disrupted as the long-wave radio communications used at the time went down for several weeks following the event.

Later that year, another major solar flare caused astronauts aboard the space shuttle to hunker down in their storm shelter, but they reported a 'burning' sensation in their eyes and seeing flashes of light even while their eyes were closed. Astronauts on the Apollo missions in the 1960s and early 1970s reported the same effect, which was caused by high-energy particles from the Sun passing through their heads and striking their retinas, causing 'sight' without their eyes being open!

According to a 2013 report by the UK's Royal Academy of Engineering, a repeat performance of the monster 1859 'space weather' incident would wreak havoc on our electricity grids, electronic and communication systems, satellites and global navigation systems. Aircraft passengers, crew and astronauts could all receive dangerous radiation doses if protective measures were not taken. The world could run into chaos if systems were down for several days. Imagine stock exchanges closed, hospitals incapacitated, logistics halted, electronic doors and lifts locked or inaccessible, shops closed as payment methods were halted. The list goes on. Pretty much everything we do would be affected, with an estimated financial impact of US$1–2 trillion and an untold cost in human life.

Given that events like this could happen on average every 100 to 200 years, it pays to be ready by monitoring the Sun's activity, responding at the first sign of trouble and building contingencies into vital national and global systems. Although humans tame nature in many ways, it pays to remember that the Sun still has the power to sweep all that aside at a moment's notice. If (or when) the next major storm comes, I think I'll play it safe and stay home for a few days. If the collapse of human society does happen,

I reckon the best thing to do will be to choose a good book and sit under the stars, reading by the brilliant purple glow of the aurora.

We should forgive our local star these occasional temper tantrums, I suppose. After all, the Sun has shone steadily upon us for 4.6 billion years and will continue to warm our beautiful blue marble until her last breath. Even when her hydrogen supply is spent and the nuclear vanishing act that powers her radiance is exhausted, she will morph and adapt to the new reality by puffing up her feathers to 200 times her current size and becoming a red giant star. As she takes on her new incarnation, she will start using the helium surrounding her core to shine. When that fuel is spent, she will slowly shed this cool orange-red layer of gas into outer space, leaving nothing but a dimly radiating star called a white dwarf.

The white dwarf stage will be the last curtain call for our Sun as a star. This small yet dignified remnant, a sun-like mass crammed into the volume of the Earth, will not shine by nuclear fusion but live off the light of her past glories. Our white dwarf sun will live on for probably tens of billions of years, like a well-loved matriarch presiding over our solar system for time immemorial.

Once the latent heat has finally been released into the cold of space, all that is left will be a tiny brown-black globule of cold gas serving as an eternal signpost to the place a once proud and luminous sun stood, giving life to trillions of creatures on planet Earth – the only place known to host life in this possibly infinite universe.

3

Red dwarfs and almost-rans

Now we've familiarised ourselves with ol' faithful, the Sun, let's meet some of our neighbours – in particular, some diminutive yet fiery characters with distinctly auburn colouring. They might be a bit of an odd bunch, with different characteristics from our star, but they are basically good eggs living out their best lives and trying (in many cases) to cultivate a family life.

Stars appear in a vast range of sizes, with most being smaller and cooler than the Sun. The smallest burn their hydrogen more slowly and therefore have much lower temperatures. You can tell a lot about a star from its colour. Hotter stars give off more blue light, and cooler stars tend to the red. It's similar to how hot gas-cooking flames are blue, whereas cooler candle flames are red. Counterintuitive, huh?

By looking in great detail at the colours of stars, we can tell precisely how hot they are, what chemicals they contain, and how old they are. We can even predict their future as they go through the ageing process.

Before we go on, I should fess up that the names of star categories make no logical sense whatsoever. From the coolest to the hottest stars we call them M, K, G, F, A, B and O, with numbered subcategories from 0 to 9. Uh-huh …

There are also subcategories for luminosity (how much light a star puts out) from 0 through to VII, in Roman numerals, just to be fancy.

Adding to the confusion, all small and average-sized stars are called 'dwarf stars'. Only the very rare and very large stars are called 'giants' or 'supergiants'. There are no 'regular' stars.

At this point I'd like to apologise on behalf of astronomers everywhere for this ridiculous situation – astronomy is littered with strange historical naming schemes. But there we have it.

Our Sun is a G2V star, somewhere in the middle of the jumbled alphabet soup of stars. B and O stars, at the top end of the scale, are humongous, gluttonous giants that explode in a shower of sparks when they come to the end of their life. In contrast, M and K stars are cool and steady types who live long and interesting lives. From now on, we will call these M and K stars 'red dwarfs'.

Red dwarfs have only about a tenth to one half the mass of the Sun, and are only half as hot. They are smaller in physical size, too, and decidedly dimmer (although not lacking in intelligence). These stars are capable of great feats of physics – they still burn hydrogen into helium to produce vast amounts of energy in their core – but because they are smaller, there are far fewer nuclear reactions and their output is decidedly more puny.

Red dwarfs are the most common type of star in the Milky Way, with the faintest M-dwarfs making up more than 70 per cent of stars in our universe. Despite that, we never see them. I mean that literally. We don't see any of them. At all. Ever. We didn't even know that red dwarfs existed before the invention of the telescope, because they are too faint to see with the naked eye. Imagine that! Every star you can see is part of the 30 per cent minority.

As slow burners, these stars live extremely long and varied lives. Not only does their reduced rate of nuclear fusion extend their life, but also their enhanced internal mixing causes fresh hydrogen fuel from the outskirts of the star to be transported to the engine in the middle. As such, we predict that red dwarfs can live in excess of a trillion years.

As adolescents they can be active and sparky. In this gregarious stage of their lives they are often characterised as 'flare' stars. Rather than being a regrettable 1970s fashion statement, this moniker actually describes a stage of intense variability in the life of lower-mass stars where they quickly erupt – explode, even – for a few minutes before relaxing back to their regular demeanour as if nothing had happened. Flares from young red dwarfs are 100 to 1000 times more energetic than when the stars are older. As red dwarfs age, they cease this nonsense and increasingly potter their way through life, with age slowly ripening and changing their character.

Stars of all sizes form when gravity pulls together materials in interstellar clouds of gas. Red dwarfs are the littlest stars, just slightly larger than Jupiter, and are the smallest member in the official category of stars. But there are also some almost-rans, plucky hopefuls who tried but just missed the cut-off for full stardom. These battlers, called brown dwarfs, live out quite different lives to other stars.

With core temperatures of below 3 million degrees Celsius, brown dwarfs are simply not big or hot enough to turn hydrogen into helium via nuclear fusion. They can, however, manage other nuclear fusion reactions involving chemicals called deuterium and lithium. These reactions don't generate as much heat as hydrogen fusion, so the surface temperatures of brown dwarfs are generally below 800 degrees. They emit almost no visible light: if you looked at one close up, you might just see a dim purple-ish glow.

Brown dwarfs weren't discovered until 1995, because the only way to find them is to use infra-red detectors. Once astronomers built a telescope powerful enough to see them, we could finally make out their slight warmth in the icy cold of space.

Let's get up close and personal with our friends making up the community of the smallest stars and the almost-rans.

Good old Ethel

Have you ever been to an ancient building and thought to yourself: ah, if only these walls could talk! What tales they would tell of ancient kings and queens, of tribal strife and everyday life before our modern world became so mechanical and disinfected.

Stories of our history are best told by the people who have lived through them. Want to know what life was like in the 1950s? Ask your grandparents or great-grandparents. Want to know what life was like 13.5 billion years ago? Ask 2MASS J18082002-5104378 B – let's call her Ethel for short – who has lived through the life of our Sun and solar system, from the times when dinosaurs roamed, through mass extinctions and almost the entire history of the universe before the Earth even existed.

Our Ethel is a diminutive red dwarf star. She weighs only one-seventh of the Sun, placing her near the absolute minimum limit of stars that are capable of hydrogen fusion. Her heart doesn't burn – rather, it fizzes. She shines so faintly that it took astronomers over a year of dedicated observing with some of the world's best telescopes to find her.

Her extraordinary life has not been spent alone, because Ethel is partnered with another ancient star – let's call her Jo. Jo is much larger, around three-quarters the mass of the Sun, but is bigger and more luminous due to her age and the way she burns.

Jo and Ethel have always been together. Just 200,000 years after the Big Bang, when the universe finally began to cool down enough to form atoms, the two stars gradually emerged from a small globule of gas that slowly shrank under the action of gravity and began spinning, flattening out like a piece of pizza dough tossed into the air. This disc of gas fragmented into two lumps that kept gathering gas, like snowballs rolling through a powdery field, until the cores of both lumps became hot enough to shine.

In those days, the only flavours of gas were hydrogen, a little bit of helium and a tiny sprinkling of lithium. These three lightest materials were generated in the raging heat of the Big Bang. Every other chemical in the universe – the stuff we're made of, like oxygen, nitrogen and carbon – was made later, inside stars. What that means is that Ethel and Jo are the most chemically pure of all stars, since they were among the very first crop ever to grow in our universe. They were rolled from the most pristine snow, before anyone trampled on it or peed on it, before any dust or pollution made it grey and sludgy.

We know exactly how pure the two stars are because we can measure the light they give off with an instrument called a spectrograph, which shows how few chemicals are in there.

As in any good long-term relationship, Ethel and Jo have grown closer together over time. As she pushed and jolted her way through the disc of gas, asteroids and possibly planets that formed around her beau, Ethel's energy was drained. When orbiting bodies lose their energy, they relax into a tighter orbit – and that's exactly what Ethel did. Today the two are close, with Ethel making one circuit around her partner every thirty-four days, locked in a universal dance.

As we invent new ways to study very faint stars like Ethel, we bring new understanding to her story of resilience and endurance. How wonderful, too, that the universal love story of the oldest stars, made up of the purest snow from the Big Bang, still echoes through the ages.

The planet killer

Imagine you're an astronomer whose passion is searching for signs of alien life in the universe. You assume that UFO myths are just that – myths – and that because little green men are unlikely to drop in on us anytime soon, we must go to them.

To search for signs of life in the stars, you seek faint radio signals from civilisations who, like us, connect via wireless radio communications across nations and continents and with a network of satellites in space. Failing that, you look for chemical fingerprints of methane and complex organic molecules in the atmospheres of other worlds that may betray the presence of simpler life forms on the surface of those distant planets. It's a big ol' sky up there – so where would you look?

A good place to start might be a planet near one of the closest stars. After all, the light from nearby stars looks brighter and we can study them in more detail. We know that the neighbourhood of the Sun is a good place for life to begin, as evidenced by us and more than 8 million other species on planet Earth.

Proxima Centauri is our celestial next-door neighbour. At 4.2 light years away, it is so close that we could theoretically fly there and back ten times within a human lifespan if we built a rocket capable of going at close to the speed of light.

But Proxima Centauri is a red dwarf star, and with only 0.1 per cent of the Sun's light output, its feeble glow is barely bright enough to serve as a night-light for your toilet, let alone to light up the vast expanses of space. Think about it – of the 1 billion trillion stars in the universe, the closest one to Earth is too faint to be seen with the naked eye. That's weird, right? It just goes to show how faint red dwarf stars can be.

Proxima Centauri appears to live in a rather delightful triplet alongside Alpha Centauri A and Alpha Centauri B. This pair are so close together that they look like a single, bright star in the sky. We know this star as the brighter of the two 'pointers' to the Southern Cross. The gravitational bond between the stars is strong, and they are close together considering that they both have approximately the mass of the Sun. They are separated by a little over the distance between Earth and Saturn.

Proxima Centauri, however, is much more loosely bound in this system, hovering at more than 1000 times further than the spacing of the larger pair. Proxima clearly isn't that into them.

Our little hero isn't completely phobic of commitment, though. It does have at least one companion – a planet with the rather fun name Proxima b. We don't know the exact size of Proxima b, but it is likely to be quite similar in size to the Earth. The big difference is in its orbit. It hugs the star with a very tight squeeze, whizzing round in just eleven days – that's one year for Proxima b (imagine the birthday cakes!). It travels so fast because it is playing a game of stellar 'chicken' with the star, orbiting roughly ten times closer than our solar system's closest planet, Mercury, circles the Sun.

In our search for life in the universe, is Proxima b the holy grail? It certainly has many things going for it. Orbiting the closest star to Earth, it is about as close as we can hope to find a planet. If humanity gets its finger out, we might just invent the technologies needed to explore this little world in the coming centuries.

The planet may be close to its host star, but remember that Proxima Centauri is a red dwarf and, as such, is much dimmer and cooler than the Sun. Proxima b lives in the so-called 'habitable zone', where any water that might exist on it could sit in liquid form. Under these conditions, life is not an impossibility. The fact that the star is a red dwarf, which can live so much longer than our Sun, also gives Proxima b longer to cultivate life. Is there any reason why we couldn't pack a suitcase and head over there right now to meet the neighbours?

Well, there are two possible flies in the ointment. First, due to its close proximity to its star, Proxima b encounters a strong 'wind' of radiation and particles streaming off the star that may strip any atmosphere the planet has. Second, Proxima Centauri is a flare star, meaning it occasionally has a dramatic outburst, as if from nowhere, rather like a solar flare. That might not be a problem, but with the planet being so close to the star *and* the occasional superflare erupting from Proxima Centauri increasing its radiation output by more than sixty-eight times, our closest planet in the hood might be getting irradiated a little too often for life to be a likely prospect.

There ain't no party like a red dwarf party

Have you ever had a neighbour who is as quiet as a mouse most of the time but once or twice a year has a raucous party? We don't necessarily mind that sort of neighbour as long as nothing gets broken, we get invited and we don't have to get the council involved.

UV Ceti is one of those types, and lives just a few doors down from the Sun in our stellar street. His house has a striking red door, denoting the M-type red giant who resides within. Nobody is sure exactly how long he has lived in the street – some say 10 billion years, but I don't think he looks a day over 5 billion.

UV Ceti is a shy star who is introverted most of the time. He likes to keep us on our toes by using a number of pseudonyms, including Luyten 726-8, Gliese 65 and LHS 10 (I like that last one). Frankly, UV is cagey about his life. Until recently, nobody knew if he had any family or friends living with him. But in 1948 or thereabouts, someone caught a glimpse of another star, BL Ceti, residing within UV's home. BL is now considered as much a part of the community as his companion.

Both UV and BL are generally good neighbours, but they are known to be a little temperamental. They say it's a medical issue: both stars have strong churning convection currents inside their bodies, which can cause strong magnetic fields to reflux around their insides (also causing the odd bout of gastro).

When magnetic fields spin, they can tangle and get wound up. Eventually, if the spinning continues, the field lines can snap. When that happens, UV and BL can quickly fly off the handle. They are not angry folks, so these outbursts manifest themselves as big, noisy and very sudden celebrations. Hot gas flies out from the source of these magnetic reconnections and giant starquakes echo from the house, reverberating from the source and causing a serious kerfuffle. Although the music can't be heard through the near-vacuum of space, it's not uncommon to see light, X-rays, radio waves and red flames shooting out of the windows. During these parties, the house can go from normal, boring, Tuesday evening levels of excitement to party central in the space of a couple of minutes. The temperature inside rises to dangerous levels too, becoming four times hotter than usual, and the house lights quickly crank up to levels 10,000 times brighter.

After a few hours, the party dies down and it seems as if nothing ever happened. The pressure valve has been released, the bottles can go in the recycling, and it will hopefully be another few months before we get a repeat performance.

Stormy weather

As we established earlier, some 'stars' are not big or hot enough to shine with their own light. As such, they don't qualify as stars at all. We call these objects brown dwarfs. They emit very little in the way of visible light and would probably appear a very dim purple if viewed up close. Due to their relatively low temperatures compared to stars – a few hundred degrees Celsius – they shine most brightly in infra-red radiation (that's heat to you and me).

One of the best-studied brown dwarfs is 2MASS J22282889–4310262, which was discovered by astronomers in 2013. It has a temperature of roughly 600 degrees and is close enough to Earth ('only' 35 light years, or 331 trillion kilometres) to allow us to see what it is made of.

By using infra-red cameras on the Spitzer and Hubble space telescopes, we can zoom in and watch as, from time to time, heat from the centre of the brown dwarf is blocked by clouds of methane and water vapour high up in its atmosphere. It is even possible to make detailed maps of the cloud patterns and winds on the brown dwarf's outer edges, using the cameras to probe the various depths and layers of cloud. In future, when we have even more powerful telescopes, we hope to use this technique to study the atmospheres of distant planets too.

Large storms are understood to rage in the atmospheres of brown dwarfs, too. These can dredge up 'rain cycles' much like we see on Earth, only in the atmospheres of brown dwarfs we don't just have a friendly Earth-like mix of nitrogen, oxygen and water vapour. In brown dwarfs like 2MASS J22282889–4310262, astronomers have found methane in the mix as well as larger particles of sand and even of iron. Particles of sand and iron are able to rise up into the atmosphere with the heat of the gas before falling back down as precipitation from the storms. Yes, folks – visit a brown dwarf and witness the incredible iron rain!

Is it a planet? Is it a brown dwarf?

Q. When is a planet not a planet?
A. When it's a sub-brown dwarf.

Scientists are well known for their love of classification. Sometimes it can help us to understand and distinguish between things, whether they are species of living creatures, different types of rocks, or stars and planets. The trouble with classifying objects, though, is that in real life, things don't fit conveniently into boxes.

We put stars into categories – sun-like stars are called G-type and red dwarfs are M- and K-type stars – but in reality there are no true boundaries, only a continuum.

Take, for example, the definitions of a brown dwarf and a planet. The International Astronomical Union (the body that decides these things) defines an object heavier than thirteen times the mass of Jupiter and capable of fusing deuterium to be a brown dwarf. A lighter object that is free-floating alone in space is called a sub-brown dwarf. Finally, an object lighter than thirteen Jupiters that is orbiting another object is defined as a planet.

All that sounds fairly simple. But in reality, the boundaries are blurred in several places.

First, some objects that are lighter than thirteen Jupiters *can* take part in large-scale nuclear fusion in their cores, because the ability to burn in this way depends on the chemical make-up of the object. Therefore, a strict boundary at thirteen Jupiters does not make sense.

Second, it is problematic that a body that is lighter than thirteen Jupiters is defined by its relationship to another body, not by its own characteristics. Although logically sound, this definition has no real physical foundation.

Take, for example, WISE 0855-0714, the coldest known sub-brown dwarf. It was formed by the gravitational collapse of a clump of gas in an interstellar cloud, just like a star. Because there was not enough material accumulated in the clump, it never made it into a fully fledged star. The temperature of its 'surface' clouds is a chilly 25 degrees Celsius below zero. It looks like a planet, it probably doesn't shine with much of its own light, and it has icy crystals suspended throughout its atmosphere. But it is not a planet since it has no star of its own.

How should we view this super-Jupiter floating alone in the cosmos? Is it really different from the same body orbiting a star? When we search for planets around other stars we find many examples of bodies with more than thirty times the mass of Jupiter.

So what is the difference between a planet and a sub-brown dwarf? If our hero WISE 0855-0714 lived with a companion, or if a nearby star captured our lonely friend as it wandered the skies, then by the official definition it would suddenly become a planet.

The moral of the story? Let's not focus too much on the definition of stars and planets. It's a rich and diverse universe out there. Enjoying the mysteries and magic of the cosmos is far more rewarding than slicing and dicing it up!

4

Cannibals

Spend long enough studying stars and you see growing evidence for their caring, sharing nature. But there are some weird and wonderful examples of 'oversharing', of cannibal stars that live like conjoined twins, in some cases completely consuming each other.

Fundamentally, this cannibalistic behaviour is driven by the universal force that drives many processes in the universe: gravity. Gravity implores stars to come close to one another and enter into stable relationships, or orbits. These are seriously long-term partnerships, lasting millions if not billions of years.

The members of stellar pairs can have many different sizes and properties. There are extremely dense and compact stars, called white dwarfs or neutron stars. Then there are regular stars like our Sun, and stupendously large giants with thin and nebulous atmospheres of hot gas. Very compact stars are more likely to cannibalise their unsuspecting neighbours, as their strong gravity empowers them to pull gas off the low-density atmospheres of their partners and literally eat them and grow in stature as a result.

Sometimes these relationships can go too far. If one member of the pair takes too much from the other, it can suck them dry. It cannibalises its partner, taking everything the other is not using, stealing its gaseous cloak and wearing it like a grotesque trophy.

This reminds me of a viral picture I once saw on social media of a possum who had apparently found its way into a bakery and eaten a tray of pastries, and was slumped in the corner in a sugar-induced stupor. As with greedy possums, this kind of behaviour in stars can backfire. If a gluttonous star can't cope with the sudden richness of the banquet it has ingested, a complete meltdown can be triggered. In extreme cases, it can cause a destructive supernova explosion that completely destroys the greedy little critter and breaks up the pair once and for all.

A stellar pairing can benefit two stars in a more balanced arrangement. Setting up a symbiotic relationship can help the parties use up more of the fuel that they normally waste. An average star might burn only 10 per cent of its hydrogen during its life, the rest simply floating back into space after the star has stopped burning. If a neighbouring star can use this gas to generate heat and light, why not?

Massive stars shine hotter and more brightly and are far more unstable and, like lottery winners and pop idols everywhere, stars that experience a rapid rise in wealth rarely stay grounded. A rise to greatness sends them off the rails. Cannibal events that pile new gas to the surface of a star often trigger the use of these massive reserves to reignite, evolve or transform.

So who are some of these stellar celebs whose new-found fame causes all manner of bad behaviour?

Is something fishy going on?

Some stars are weird – I think we've established that. But some are so weird that we're not even sure we can understand their life story.

The star BP Piscium in the constellation of Pisces (the fishes) is one such enigma.

BP Piscium is reddish in colour and its equator is surrounded by a 'spare tyre' – a disc of gas and dusty particles that has flattened out as it rotates around the star. BP Piscium also spews out two narrow jets of gas from its north and south polar regions, which act as gigantic interstellar markers to its position in space.

At first glance, astronomers thought that BP Piscium was a very young star, still building itself and its solar system of planets much like the Sun did 4.5 billion years ago. On closer inspection, though, using sophisticated instruments on ground-based and space telescopes, it became clear that this was no ordinary juvenile star. What tipped astronomers off?

A few things just didn't add up. The way that larger and smaller particles were distributed throughout the disc wasn't as neat and tidy as we would expect in a young star, and the amount of the chemical lithium in it was seven times less than would be expected for a youthful star.

The current scientific view is that BP Piscium is actually much older and wiser than we first thought. It seems to be an evolved star, older than our Sun and puffed up by the extra heating of a shell of hydrogen around its (mainly) helium core. A red giant star enveloped by a cloud of dust, so much so that most of its light is blocked. A star that shines 75 per cent of its light in infra-red.

This dusty shroud may be hiding a sinister secret.

As highly evolved objects, red giant stars have no business hanging around with a disc. It's like a 75-year-old swanning down the street with a hula hoop – fine for an eight-year-old, you might think, but unusual for a senior citizen.

Some astronomers have suggested that the swirling disc of gas may have been caused by a stellar cannibal that has consumed a nearby star. Others argue that there is no direct evidence for this swallowed ex-star and that further work is needed to prove the case one way or the other.

Recent studies of the disc using infra-red telescopes show evidence of a huge planet and possibly a swarm of comets that may be forming from the material released by the massive disruption. A star dies, and a solar system is born.

Whatever the dark history of BP Piscium, we can be certain it is a fascinating enigma. The fact that it may be hiding a secret past of treacherous cannibalism only adds to its appeal.

KIC 9832227

How about this for a steamy salsa dance? KIC 9832227 is made up of a wild pair of stars twirling around each other so closely that they are practically joined at the hip.

In science-speak, KIC 9832227 is an eclipsing binary star system. In other words, two stars (the binary) eclipse one another on a regular basis (every eleven hours, in this case) when one passes in front of the other (as seen from Earth). That is useful because, as with most binary stars, they are too far away to be seen as two separate objects with even the most powerful telescope.

Even though they can't be seen independently of one another, we use the behaviour of this pair of stars playing hide-and-seek, eclipsing each other, to study the brightness and colours of each individual, their masses and their distance apart. They really do orbit very closely, with a separation of only three times the radius of the Sun. Neither star is particularly unusual, but their proximity to each other means they actually *share an atmosphere*, giving the system a peanut shape.

That's why KIC 9832227 is also known as a contact binary – in other words, the stars are so close that their atmospheres are touching. Over time, the distance between them will get smaller and smaller and they may eventually merge into one. The result will be (for a few days at least) one of the brightest stars in the sky – a luminous red nova. In astronomy, 'nova' means 'new star' – and such an event in our own Galaxy would be a rare and exciting sight.

We have witnessed similar coalescence of stars before.

In 2008 a 'new' star was discovered in the constellation of Scorpius. Although not bright enough to be noticed by members of the public, it was certainly obvious to astronomers. By looking back at old pictures of the stars, astronomers discovered that the 'new' star was in exactly the same place as a pair that had been orbiting each other so closely that they were actually touching. Sound familiar?

Digging out data from the previous few years, astronomers saw that the pair of stars had been orbiting each other every 1.4 days, a period that started to get exponentially shorter and shorter until the moment of the outburst. At that point, the star brightened by 10,000 times. Kaboom! A luminous red nova.

The merging of these two stars was the first in human history to be witnessed at such a level of detail. Let's hope that KIC 9832227 treats us to a similar show sometime soon. Since it is only 1900 light years away from Earth, it will be far more spectacular!

Making a splash

When two stars merge like our mate KIC 9832227, we find out from the self-indulgent displays of a bright 'new' star in the sky: 'Look at me! – I'm a new star!' But there is another secret signal from such tumultuous events that human beings have only just mastered the art of detecting, a signal far more subtle, but powerful; the encrypted messages of the universe that can show us collisions between stars so far away that we might never detect their light.

Gravitational waves

What are gravitational waves? They are disturbances in space and time. They are like the swell on the ocean, the ripple in freshly washed bed sheets as you cast them across the room like a fishing net, the stretching and squashing of the earth beneath your feet during an earthquake.

Gravitational waves do not travel through space, like light, and are not like sound waves, which depend on the motion of particles through the air. They are ripples in the very fabric of the universe. We call this spacetime.

Spacetime is an idea from the general theory of relativity, which Albert Einstein conjured up more than 100 years ago. It is the imaginary surface – like a stretchy sheet of plastic – that helps us imagine how stars and galaxies move in space. This concept helps us picture how the gravity of large objects such as stars and planets causes space and time to curve.

Empty space is like a backyard swimming pool, full of calm. You sit alone and still, contemplating the beauty of a summer's day. Small leaves fall gently from surrounding trees, spinning and dancing down to float gracefully on the water. A pond skater sits nonchalantly on the glassy surface, taking in the sounds of birdsong. Sunlight reflects from the water, which is like a polished mirror, and kisses your skin. As if from nowhere, your two kids jump in, closely followed by a labrador (and you don't even have a dog). All hell breaks loose as the children link arms and spin around, generating a wave that ripples quickly to the edge and splashes your face and your sunglasses and ruins the book you were reading.

Spacetime is your swimming pool, and the gentle, calm motions of the water represent the ordinary paths of stars, planets and galaxies gliding elegantly through space. The kids? A pair of super-heavy stars in a death spiral and merging together. Splashing and twirling, they make gravitational waves as they orbit closer and closer before BANG! – the waves end abruptly as the stars merge.

A merger of two unusually dense stars – neutron stars – happened on 25 August 2017. At least, that's when the gravitational waves were detected by the Laser Interferometer Gravitational-Wave

Observatory (LIGO) in the USA in an event called GW170817. The stellar collision happened 140 million years ago, but the gravitational waves (travelling at the speed of light) took all that time to reach us. That's because the colliding stars were in a very distant galaxy called NGC 4993, which lies 140 million light years away from Earth. At the same time, a spray of extremely energetic gamma rays was seen from the galaxy.

Coincidence? No way. Gamma rays are only generated in these numbers by ultra-high-energy events. The catastrophic destruction of two stars definitely qualifies as one of those.

Nasty 1

Nasty 1 is a cosmic cannibal. Its strange name comes from a rather benign place. It's the first member of a star catalogue assembled by J.J. Nassau and C.B. Stephenson in 1963 listing 'luminous stars in the northern Milky Way'. They were discovered using grand old telescopes in Hamburg and Cleveland, Ohio, and the first two letters of the surname of each astronomer give us – ta-da! – the NaSt catalogue.

Nasty 1 has some skeletons in its closet. It has all the characteristics (e.g. colour and temperature) of a Wolf-Rayet, a massive star that has finished burning steadily and is now rapidly throwing off its outer layers of gas as the hot core goes through a series of dramatic changes at the end of its life. From what we can tell, Nasty 1 may be caught in a rare, changing moment in its life as a very massive star.

Usually, in such a star, we would see a fiercely hot helium-burning core surrounded by a fluffy atmosphere of chemicals previously produced in this core, including carbon, nitrogen and oxygen, that are escaping rapidly from the star's clutches. Normally the hot plasma would puff out in a sort of 'dumbbell' shape as it is guided into space by the magnetic field of the star. But when astronomers zoomed in to study Nasty 1's shape they found that instead of a fluffy atmosphere, it was actually surrounded by a huge dusty disc of material orbiting around its middle.

This is not normal. Old, hot stars don't do discs.

So how did the disc get there?

It's a little hard to tell since the dust is so thick, but we think it may be being shaped by an unseen companion, a tiny twin star that is snacking on the gas around the central giant. As it orbits around the giant, the unseen companion's gravitational field pulls the giant's gas cloak into a disc, like an unseen hand moulding a lump of wet clay on a spinning potter's wheel. Not just shaping the gas, the invisible mate is gorging itself on the bigger star, pulling its atmospheric gas onto its own surface. It's a stellar cannibal.

We think that it may suffer from its current feast, since stars often can't cope with a sudden increase in mass. It could cause instabilities or even explosions in the smaller star for (thousands of) years to come.

So what future for this odd pair? As the cannibal star grows and cleans up the bigger star's atmosphere, the disc will likely disappear. This will allow future astronomers, perhaps in a few tens of thousands of years, to get a clear view of the strange couple within.

For the giant star, an even more cataclysmic future is likely. Such stars often end in a supernova – a complete collapse of the core of the star followed by an obliteration by explosion. A calamitous divorce indeed.

Blue stragglers

In the far reaches of our Galaxy lie gigantic stellar clusters, buzzing swarms of stars, some numbering upwards of a million members. We call these groupings 'globular clusters'. Their centres are the most crowded places in our universe, with two stars packed into every cubic light year. (Doesn't sound like much, but that's 500 times more squashy than the stars in our neck of the woods.)

Globular clusters are the retirement communities of the Milky Way. They are almost exclusively made up of very old, cool red stars. Most of these stars have lived for more than 12 billion years, almost the entire age of the Galaxy, and they have earned a bit of down time.

When we study the clusters, something puzzling stands out: there are always a few hundred hot, bright blue stars mixed in that would appear to be a few billion years younger and a little bulkier than the rest of the population. It's kinda like finding a bunch of chunky toddlers haring around a retirement village.

How did these young, blue stars get into a cluster that formed more than 10 billion years ago? Did they form there recently, or fly in from the outside?

Never fear: the sleuths with the telescopes have figured out what's going on.

In normal parts of the Galaxy we would never see two independent stars approaching one another, with the exception of stars that have formed from the same cloud and live happily together for billions of years, paired in a binary orbit by the force of gravity. But in the crowded centres of globular clusters, stars jostle for territory. In the stellar equivalent of Tokyo's Shinjuku railway station, stars crush and heave through the crowds and at times actually come to blows.

That can happen in two ways. One is for a pair of stars to collide and quickly merge under the influence of gravity. The other, slightly more convoluted, is that a pair of stars come close together and are captured in a binary orbit around one another by the action of gravity. The pair become so close that gas from the atmosphere of one star is pulled onto the other.

When either of these scenarios plays out, fresh and unadulterated hydrogen gas from the outside of one star falls onto the second star, bringing a new source of fuel to stoke its fusion burning. This quickly leads to a 'rebirth', with hotter and faster nuclear fusion flaring up in the core of the star that is receiving the gas. It's a bit like chucking another log on the fire or throwing petrol onto a flame: you get a big, hot reaction.

Blue stragglers are not only found in the centres of globular clusters. They were recently discovered in the crowded regions towards the centre of the Milky Way.

Wherever stars are trapped in close proximity, it seems, some see fit to share their resources and only get stronger for it. A lesson for us all, perhaps?

Families

Ah, families.

Despite the rise in the number of people living by themselves, humans around the world still live predominantly in families or groups. The 2016 Australian census showed that 76 per cent of Australians lived in multiply occupied households. In the UK, 70 per cent live with at least one other person. In contrast, many people in African and Central Asian countries live in larger clans, with the average number per household being typically four to five people.

Family relationships can be complex and are sometimes messy, but for many of us they are the bedrock of our lives.

Stars often form families or groups, too, linked by their mutual gravitational attraction. An estimated 85 per cent of stars, like our Sun, live with others. What seals the fate of these stars as sociable creatures is the circumstances into which they are born. Their genesis comes in turbulent clouds of interstellar gas and dust that collapse under the action of gravity. They fragment into distinct clumps and then gravity does the rest, slowly moulding a pair or group of stars that live together forever, whether they like it or not.

For smaller stars the story may be quite different. Recent studies of red dwarfs suggest that these dim globules are predominantly loners, with about 75 per cent staying happily single for their trillion-year-plus lifespan. Why could this be?

These diminutive stars probably form in smaller interstellar clouds than larger stars do – and smaller clouds are less turbulent. That leads to less lumpy gas and therefore a smoother, more stable accumulation of material. When gravity does its assembling job, the ultimate result is a crop of predominantly single stars.

As soon as human beings began looking through telescopes, we realised just how many of the bright stars that are visible every night are actually made up of two, three or more stars that are simply too far away for our eyes to separate them. Our celestial neighbours love to find a life partner, and some even have extended families.

What would it be like to live on a planet with multiple stars?

For starters, sleep might be a little tricky. With several suns you would be treated to great sunsets (and plenty of them) but almost certainly have to contend with longer days and shorter nights as the sky juggles with its cacophony of brilliant orbs.

Do these incredible places exist? They certainly do. And there are extended families with two or more stars, some with a clutch of planets for good measure.

Like *Star Wars*, but real life

If you've ever seen a *Star Wars* film, you might remember that the hero Luke Skywalker comes from a planet called Tatooine.

At first glance this (fictional) alien world appears much like the Earth, with deserts, mountains and vast plains. Look up in the sky, though, and instead of a single sun, Tatooine is blessed with a pair of orange-red stars casting their light across the desert surface. From the perspective of its inhabitants this generates a beautiful twin sunset that serves as an evocative backdrop to memorable scenes in the movies.

Is this figment of George Lucas's fertile sci-fi imagination scientifically credible?

On the whole, yes. We know that stable planetary orbits are possible around double stars, and we have now discovered examples of solar systems *in real life* that are even more incredible than the one in the *Star Wars* universe.

Take, for example, LTT 1445 ABC, a dazzling trio of real stars – well, dim red dwarfs – that form their own solar system lying just 22 light years away from our own. The brightest star (LTT 1445 A) is a faint orange globule a little over a quarter of the size of our Sun. It is joined in a pretty dance by two slightly smaller red dwarfs, called B and C, that are spinning in a mutual orbit around star A. These two stars are slightly smaller and fainter and much cooler than the Sun, with surface temperatures hovering around 3000 degrees Celsius (compared to 5800 degrees on the Sun).

Also orbiting the brightest star is a rocky planet called LTT 1445 Ab. With an estimated radius of around 6000 kilometres (give or take), it initially seems very Earth-like. However, LTT 1445 Ab is orbiting a red dwarf star that is far cooler than the Sun, so you might think it is cooler than Earth.

You'd be wrong.

LTT 1445 Ab zips around the star every five days or so in an extremely tight orbit less than one-tenth the distance between our Sun and her closest planet, Mercury. This means the brave little planet is being bathed in five times more heat and light than Earth receives from the Sun. From what we can tell, LTT 1445 Ab swelters in estimated temperatures of 160 degrees Celsius, far above the boiling point of water.

What would it be like to live on this scalding rock?

Jennifer Winters, an astronomer at the Harvard–Smithsonian Center for Astrophysics, studied the planet and imagined the scene.

She said, 'Standing on the surface you'd see one big orange sun and two much smaller orangey-red suns in the distance. The [main] star would look really big in the sky. It's really close. The other two are much farther away. They'd look about 100 times brighter than Venus, and about the same size in the sky.'

Not a great place to live for us due to its soaring temperatures, but at least we'd enjoy a nice sunrise.

Astronomers are excited about the prospect of learning more about this planet thanks to a chance alignment in the heavens between the star, the planet and Earth that means that once in every orbit, the planet passes in front of the disc of the star, a bit like a solar eclipse. That's amazing because it could allow us to start studying the composition of the planet's atmosphere, if indeed it has one. With a powerful enough telescope, we might be able to search for telltale signals of gases that enable life to flourish or that are waste gases from living creatures. If oxygen, carbon dioxide, water vapour or methane is present in the planet's atmosphere, we will see certain colours 'nibbled' out of the star's light by absorption in the atmosphere of the planet. This could help us to answer the question: is the planet habitable?

At present it's an open question, especially given the high temperature of the planet. There is hope, however. Certain types of life called 'extremophiles' exist on Earth that can survive under extreme conditions of temperature, pressure and acidity, hiding out at the bottom of deep ocean trenches and within the calderas of active volcanoes. It is not outside the realms of possibility that life has developed under what we consider to be 'extreme' conditions. We just have to keep looking.

I'm looking forward to a time when these atmospheric studies are possible with the next generation of telescopes. How exciting to find out more about the likelihood of alien life existing in this awesome solar system close to our own!

The four-star planet

You might think you need to be a highly trained astrophysicist with a giant telescope and a keen grasp of complex orbital mechanics to discover a new planet. Think again.

In 2013 two amateur astronomers, sitting comfortably in their homes, used data from NASA's Kepler mission to uncover KIC 4862625, a large Neptune-sized planet with a substantial extended family all sharing a single living space.

KIC 4862625 was found via the planethunters.org project, which enables members of the public to scrutinise data from the Kepler Space Telescope. Participants scroll through images of 'light curves' (graphs of the brightness of stars taken continuously for days, weeks and months) to identify signs of planets that are eclipsing their parent stars and temporarily blocking some of their light.

The discovery was made by 53-year-old computer executive Kian Jek from San Francisco and 68-year-old oncologist Robert

Gagliano from Arizona, who both noticed small dips in the light from the stars in this region. They flagged their suspicions on the website, which alerted a team of professional astronomers who then went about calculating the mass (20–55 times Earth) and diameter (6.2 times Earth) of the new planet.

With further study it was found that the planet resides in a complex star system, Kepler-64, that contains four members. Somewhat unusually, it is performing one big circle around two of the stars. How does that work in practice?

Imagine a pair of stars, one similar to our Sun and the other a red dwarf. Two parents in a committed relationship who orbit each other once every twenty days. Outside this inner sanctum is their child (the Neptune-like planet discovered by the plucky citizen scientists), who circles her parents once every 138 days. We call this a circumbinary orbit, and it's a big deal.

Most planetary discs form around a single star, even if the star itself lives in a pair. Not this time. KIC 4862625 almost certainly formed from a single spinning disc of gas and dust that formed around both stars about 2 billion years ago.

That's not the full story, however. Even further out (the equivalent distance from the Sun to the edge of our solar system) there is another pair of stars, one sun-like and the other a red dwarf. We might think of them as an aunt and uncle of the inner pair. These extended family members may have formed within the same interstellar cloud, but they didn't give birth to the planet.

The four stars may share a home, but somehow they all manage to get along. Gravity acts as a constant shepherd, keeping the stars and planet in a stable configuration, adjusting and moving each member until it hits its stride. Like a highly trained motorcycle stunt troupe, the members of the family keep a steady rhythm and manage to avoid conflict.

As more and more complex families of stars like this are found, I think it is brilliant that citizen scientists are playing an important role in the discoveries. Why not sign up for one of these projects? Next time it might be *you* uncovering an alien world!

Five-star review

Contact binaries are always touching. So close are these star pairs that their atmospheres are continually mixed together like a stream of hot milk poured into a rapidly stirred cup of coffee.

Contact binaries are the result of stars living in close proximity for long periods of time. The universal force of gravity pulls them together – including the gas – and causes their atmospheres to mix.

In 2013 a team of astronomers announced the discovery of a love-struck pair of stars who live within a complex stellar quintuplet. Their rather snazzy name is 1SWASP J093010.78+533859.5.

The five members of this family are the touching pair, Laura and Dan, who can't stop whispering and giggling and cuddling; a long-term married couple, Ed and Pete, who are stable but have drifted apart somewhat (3 million light years, to be precise); and bright single Maria, who loves her independence but lives close to Ed and Pete and socialises with them regularly.

Laura and Dan are never apart. These love-locked stars are like a newlywed couple whose sweaty palms are permanently clasped together. As you can imagine, their friends and family find this delightful and not the least bit annoying. The saccharine duo orbit each other every five-and-a-half days in a never-ending slow dance that sees their very beings mingle.

On the other side of the family, Ed and Pete have a healthier relationship, enjoying their own space and their own personalities. They do love Laura and Dan, but find they don't have much in common. Maria, the single sister, largely does as she pleases, yet gravitates to the boys for company.

The whole family is on a spinning carousel the size of our solar system, with the two branches of the clan remaining safely on opposite sides of the orbit, around 21 billion kilometres apart. Their distance keeps them safe, since proximity in the face of gravity can lead to unstable orbits. Clashes can arise. Stars can even come to blows, with devastating consequences such as stellar cannibalism and supernova explosions.

These family members remember to send holiday and birthday greetings but commit to little more. They certainly dare not get together for Christmas dinner in case they all have an argument and dissolve into a single, gigantic black hole.

Castor – six stars in one

Ever heard of Gemini? Of course you have – it's on the horoscope page in all the magazines (don't get me started on that).

To astronomers, Gemini is the ancient Greek constellation of 'the twins', so-called because its two brightest stars make a very handsome pair. They were reputed by the ancient Greeks to be twin half-brothers, born from different fathers and hatching from an egg of their mother, Leda, who appeared in the world as a swan. Families can be so complicated.

Anyway, the brighter star in the twins is Pollux, a gentleman giant shining with an orangey hue. The fainter brother is Castor, a star who is all grown up and has formed a very complicated family of his own.

When you look at Castor with the unaided eye he appears as one fairly bright, bluish-white star. With a small telescope you can separate him into *two* fairly bright bluish-white stars. When astronomers in the 18th century found that Castor was actually a double, the two stars were named Castor A and Castor B. Cool!

Later, some smart alec came up with an even fancier telescope and found that one of the pair, Castor A, was itself a double star. They quickly renamed Castor A (the bluish-white one) Castor Aa and the other (a red dwarf) Castor Ab. Confused yet?

You'll never guess what happened next.

Castor B was found to be a double star too, made up of a bright bluish-white one and a faint red dwarf, just like Castor A. The renaming committee came back out and gave them the monikers ... you've guessed it: Castor Ba and Castor Bb.

Despite how migraine-inducingly complicated that all sounds, the Castor A double star (Aa and Ab) and the Castor B double star (Ba and Bb) actually make a neat little pair – a double-double, if you like. They orbit one another's centre of gravity once every 400 years or so, which means you can see their positions move within a human lifetime.

A while later, Castor C was discovered. This faint red star, ambling around in a 14,000-year egg-shaped orbit, is not visible without a telescope. When probed further using a special camera that measures the exact brightness of stars, Castor C was discovered to experience regular periods of dimming every few hours before it returned to its original brightness. This happened like clockwork, and astronomers quickly figured out that the dimming was caused by another body moving across the front of Castor C. With a bit more digging it was confirmed that Castor C is *also* made up of two stars. These two red dwarfs in a close binary pair were renamed Castor Ca and Castor Cb.

A little recap: the fainter twin of Gemini is actually made up of a double-double star, with each pair circling each other; and the pairs, in turn, are circled by *another* double star. Sextuplets in space. Simple!

The magnificent seven

When I was at primary school in the pretty English village of Finchingfield, I was given the opportunity (OK, forced) to learn a charming traditional country pursuit called maypole dancing. My school chums and I would practise for weeks on end, preparing like elite athletes to perform to adoring crowds (all right, our parents) at the much-awaited May Day village fair.

Maypole dancing consists of a large number of children, dressed in white, dancing around a vertical 2.5-metre metal pole in a bizarre medieval ritual that somehow survived the space age. Each child holds a piece of coloured ribbon that has been attached to the top of the pole and, to the strains of upbeat English folk music, prances around the maypole in unison, circling carefully as you would in a traffic roundabout.

The violin, accordion and drum would beat to a giddy tempo as we skipped about, dashing over then under the ribbons of the children on the inside lane. Suddenly, at the eighth beat, half the children performed a brisk about-face and we now had *two* streams of kids flying around the maypole, precariously ducking and weaving in different directions, spinning and tangling their ribbons in a flurry of English countryside goodness. The teacher, Mrs Barker, barked instructions: 'Clockwise, boys!' 'Under then over ... that's it ... now TURN!' 'Girls, smile. Now anticlockwise. Over, under ... now pass the ribbon to your other hand!'

The outcome of this bizarre ritual was the spinning of a complex silken web of red, blue, green and yellow. Some parts of it looked smart, with neat lines like a Windsor-knotted tie; other parts were tangled and uneven (those were my bits, as I remember). The whole spectacle made me feel like my classmates and I were pawns of an all-knowing arachnid queen.

Where did my forebears get their inspiration for this complicated ritual? From weaving, perhaps. Or even from space.

In the constellation of the Scorpion is a faint star called Nu Scorpii, or Jabbah (its Arabic name). With a large telescope we find that this star is actually a complex stellar system with no less than seven siblings. This band of dancers spin around their solar system in a perfectly choreographed display that has lasted thousands of millions of years. Imagine the colourful web their ribbons would weave.

Large groups of stars bound by gravity are complex and unstable, so by their nature they are less likely to survive for long periods of time. As they tumble and spin they get out of step, which can lead to cannibalism – or even exorcism, as errant stars are flung from the system. But survive they can, as Jabbah has done for many millions of years.

Envisaging this complex family of seven siblings is difficult. There is no big parent star at the centre of things. Instead, there are two cliques of stars in the Jabbah system, and the two clans live on each side of a giant orbit shared by the magnificent seven.

One side of the system has three stars. Sister 'C' (Claire) and a pair called 'Da' (Darren) and 'Db' (Deb) spin in time about the centre of the family. If they were holding threads, they'd be twisting and twirling a pretty impressive tartan in their own right. On the other side of the orbit sit the other four, performing their own silent ballet. Brother 'B' (Brian) circles closest to the centre, with 'Ac' (Alec) completing complex pirouettes around him and Alec himself being orbited by a pair, 'Aa' (Ayla) and 'Ab' (Abi), as we move furthest from the middle of this melee.

The motions of this star system look like the flight paths of a squadron of 747s circling above a fog-bound airport.

Is this sort of thing typical? Not at all. Our friend Jabbah is one of only two known septuple star systems in the universe, although there are sure to be many more that we simply haven't discovered yet.

Jabbah's interstellar maypole dance proves to us that examples of extreme complexity can readily exist in the universe. Can you imagine living on a planet within the Jabbah 7 stellar system? Seven stars in the sky. Night-time would be all but absent, with perpetual twilight even if all the stars were momentarily below the horizon. There is no such thing as a dark starry sky.

This thought really makes me appreciate the beautiful simplicity of our own solar system, with a single star and eight planets orbiting in near-circles. A single sunrise and sunset. Day and night, light and dark. A magnificent blanket of stars.

Thank you, Sun.

Enigmas

To our ancestors, the sky at night was a fertile source of knowledge about the weather and the seasons, and for direction finding. The stars were the 'heavens' and the 'gods'. Constellations were constructed as a means to explain the world, to impart moral stories and keep the population in line.

Astronomical science has been around for tens of thousands of years. Since the advent of powerful telescopes, though, humans have learned more about the physical nature of the stars themselves. How their colours and the secret fingerprints in their light reveal their chemical make-up and their temperature. How the invisible particles we receive from the Sun called cosmic rays explain how they shine. We follow the trials and tribulations of their lives as they are born, evolve and die just like we do.

Some stars elude understanding. Despite all the millions of dollars of astronomical kit thrown at these stars, the best minds in the world still can't figure out what's going on. It is these stars that perhaps deserve the most attention – the oddballs, the enigmas, the ones that elude all understanding, that refuse to live by the rules and sit in the neat and convenient boxes that we scientists like to build for them.

There are many people who approach life in such unpredictable ways, who carve their own furrow and take brilliant short cuts nobody else can see. Or who follow long, winding paths far away from the roads most travelled and discover great things along the way. We call these people creative, brilliant – leaders, even. Some are described as geniuses. But many who take a different path are smacked down, discouraged or even persecuted. In human society, difference is celebrated and punished in equal measure.

Often, though, we can learn the most from those who stand out, the great inventors, innovators, writers and artists who defy convention and throw themselves into their calling. From Marie Curie to Michelangelo, Grace Hopper to Galileo – these thinkers and creators move far beyond the narrow paths that constrain so many of us, leading to stupendous outcomes.

Stars are not so different. Some are brilliant, making their mark beyond the shackles of the Earth. They make their own astronomical story, baffle us with their secrets, or shine in a different way. Studying such interstellar enigmas teaches us more about the wild possibilities in the universe. These rogue stars do more than those we understand to help us discover the secret lives of the stars.

Dusty star or aliens?

Look at a photo of the constellation of Cygnus (the Swan) and you probably won't even notice Tabby's star, an unremarkable dwarf star 1470 light years from Earth with a size and temperature similar to the Sun. Watch its behaviour for a few weeks or months, though, and it reveals an astronomical mystery like no other.

Also known as KIC 8462852 or Boyajian's star, Tabby's star is named after astrophysicist Tabetha Boyajian, who discovered its very strange behaviour in 2015.

Tabby's star shines fairly steadily for weeks or months on end but then suddenly dims by up to one-fifth of its usual brightness, as if 'power save' mode has engaged. The dimming episodes are neither regular nor predictable, with some dimming the star by only 1 per cent and others by as much as 22 per cent.

There are a few possible explanations for a star changing like this. One is that the star itself is changing its light output. Another is that something is passing in front of it, blocking some of its light. Studies in 2016 and 2017 suggested that the process by which heat bubbles up to the surface of Tabby's star sometimes breaks down for a time. This could potentially cause dimming.

Many scientists prefer the explanation that some object, or objects, are passing in front of Tabby's star. Because the dips in its light output are so large and irregular, the object cannot be something simple like a planet or brown dwarf star. Suggestions that the light is blocked by a large thick disc of gas and dust – like the disc that forms planets, or the debris from a planetary collision – don't stack up, either. Such a structure would be easily visible since it would shine brightly in infra-red (no such emission is seen).

One team of scientists came up with the idea that perhaps a gigantic swarm of comets (which usually live far away from a star) has been disrupted by some gravitational calamity and is now orbiting close to Tabby's star. Due to the heat from the star, the comets are melting and breaking up, causing clumps of dust that periodically block the light.

Unfortunately, we would need a gigantic number of comets – 650,000 comets, each 200 kilometres wide – to cause the observed dimming, so this idea too has been discarded.

Perhaps something more solid is blocking the light?

Some scientists have toyed with the idea (albeit in a fairly tongue-in-cheek way) that an intelligent alien species might have constructed a gigantic 'Death Star'-like megastructure made up of (say) millions of solar panels that is harvesting energy from the star for use by their civilisation. Back-of-the-envelope calculations revealed that such a structure would need to cover about 750 billion square kilometres to cause the observed fading of the star's light. That's 1500 times the surface area of Earth. Follow-up searches by the SETI (Search for ExtraTerrestrial Intelligence) Institute for radio signals or laser light from alien construction sites gave a negative result.

Thanks to several years of frenetic study with telescopes around the world, we are getting closer to understanding this weird and wonderful star. Almost 20,000 separate observations have been made over the past few years using optical and infra-red cameras, and these measurements have revealed that whenever Tabby's star dims, blue light from the star is blocked more than red and infra-red light.

This proves that whatever is blocking the light *cannot be a solid object* like a planet or an alien structure. The only substance we know that allows infra-red light to preferentially pass through it is dust (not the yucky human-skin dust that you get at home, but what we call particles that are bigger than atoms floating in space). The light colour profile tells us that whatever is causing the star to fade *must* include particles of dust that are 1000 times smaller than the width of a human hair. This revelation in itself presents more fascinating questions.

How do large pockets of dust come to be around a bright, older star? Radiation from a star like Tabby's would sweep away small grains of dust within a matter of months, like a leaf blower on an autumn day. There must be some process by which the dust causing Tabby's dimming is either protected or replenished.

Calculations show that the dimming events we observe would require about one Moon's worth of dust. Does the dust come from a moon or a small planet from Tabby's solar system being slowly evaporated by the intense heat of the star? Or is an alien simply recharging her planetary batteries?

Eventually we might get to the bottom of the secrets of Tabby's star, but for now she remains the queen of mystery.

The stars that disappeared

Scientists have started putting in significant effort in recent years to study not only *what we can see* in space but also to detect *things that change* in the universe. Sometimes the most puzzling stars – stars that have brightened, dimmed or moved quickly through space – are uncovered by searches such as these.

With the advent of modern computing techniques able to identify bright objects in pictures of the stars and compare images automatically, the task of identifying these changes has become a lot easier.

In 2019, a team of researchers compared images of large areas of the sky comprising 600 million objects and spanning around fifty years. Using old US Navy star catalogues alongside images taken using modern telescopes, they found that more than 150,000 stars appear to have vanished since their neighbourhood was last photographed.

Looking more closely at the images, the team found that many disappearing star candidates were based on poor-quality older photographs or were stars that had simply moved. They discarded these images and were left with just 100 examples of high-quality images showing stars, shining predominantly in red light, that had genuinely appeared or vanished.

What happened to them?

We don't yet know, but they could be nearby red dwarfs that have flared and brightened by thousands of times. They could be distant hot blue stars going supernova (exploding) whose light has been stretched in wavelength as it travels through the expanding universe and now appears red. They could be 'failed supernovae', or massive stars that have somehow skipped the cataclysmic explosion that ails most of their peers and gone straight for the dive into their own bodies to form a black hole. Maybe they have gone to the same place as all the odd socks in my washing machine.

Even more exotic (albeit unlikely) explanations abound. The appearing or disappearing stars could be due to alien megastructures (e.g. spheres made up of solar panels) built around the stars to harness power, or lasers shone across the Galaxy as a means of interstellar communication by technologically capable extraterrestrials. I won't be betting my house on either of these two explanations, but in the absence of other information you never can be sure.

Whatever the answer to the mystery, these stars are now a rich dataset for further study. Astronomers are rolling up their sleeves to find clues to the whereabouts of the disappearing stars.

This is epic!

Space is epic. Want to know what else is epic?

EPIC 204376071.

With only 16 per cent of the Sun's mass and 3 per cent of the Sun's shining power, this young and compact red dwarf star in the constellation of the Scorpion glows weakly, like the 20-watt lightbulb that barely raised a glow in our family's living room when I was growing up.

What makes EPIC 204376071 so compelling is that the star has been caught playing a game of interstellar hide-and-seek. Rather like Tabby's star, every now and again EPIC 204376071 experiences a monumental dimming event. Whereas Tabby's star sometimes dims suddenly by more than one-fifth of its usual brightness, our awesome little friend has been known to lose up to *four-fifths* of its light output in a matter of hours before slowly returning to its original shiny self. It's as if the star is teasing us with an exaggerated display of coyness.

What sort of event could cause this diminutive body to fade so dramatically? Measurements of how bright the star is over a long period of time have given us some very strong clues.

EPIC 204376071 has been observed extensively by astronomers using the Kepler Space Telescope, who racked up a total of 160 days of observing over two separate years. They found that the red dwarf experiences frequent flares, where the star emits a glob of hot gas and briefly becomes brighter. It also suffers from acne, with groups of sunspots coming into view as it spins once on its axis every 1.6 days and causing the light from the star to change by just over 1 per cent.

The big event, though, was a one-off – captured by chance in 2017 when EPIC 204376071 rapidly faded to less than 20 per cent of its original brightness and then took twice as long to recover to its previous level as it did to fade in the first place.

This was the smoking gun. The depth and asymmetry of the eclipse proved that the event was not caused by another star or planet passing in front of the red dwarf. An orbiting star or planet cannot feasibly block 80 per cent of the light of a star. Think about it: an eclipse caused by a spherical object such as a star or planet could never take twice as long to end as it did to begin. Unless the object was huge and teardrop shaped, a solid body eclipse seems unlikely.

Another argument against a star or planet is that an (unseen) star passing in front of EPIC 204376071 would reduce the light to no less than 50 per cent of the total, since it would also contribute light of its own to the picture even when the pair were in full eclipse. If the eclipsing star were very dark (e.g. a brown dwarf), it would be small and therefore not block all the background light from the hidden star. Either way, it's an implausible explanation for the 80 per cent dimming that was observed. Similarly, no planet is large enough to cover 80 per cent of EPIC 204376071 – especially with an orbital period of more than 160 days (proving that the planet, if it did exist, must be very distant from the star).

We know that a star disappeared for a day and something passed in front of it – but what?

As with Tabby's star, we are pretty sure it has something to do with space dust. Measurements taken with infra-red cameras show that the star is probably surrounded by a disc or cloud of small interstellar debris.

Scientists who observed EPIC 204376071's weird disappearing act have come up with a list of possible explanations. They tested two scenarios by re-creating each in a computer (we call these re-enactments 'models'): (1) the star is being circled by a large planet that has a dusty ring (like Saturn's), and (2) a clump of dust from outside the star fell onto its surface, causing dramatic fading of its light output. They found that either scenario was plausible, but one was more likely. The first model – a planet with three times the mass of Jupiter, surrounded by a ring of dust four times the width of the star – exactly recreated the observed dimming event. With the second model, a sheet of dust in a permanent orbit also produced good results, but was a less convincing match to the observed eclipse.

Given that the Kepler Space Telescope is no longer operational, it may be years before we observe another epic dimming event from this fascinating star. Until then, we will just have to use our imagination!

Przybylski's star

What Przybylski's star lacks in vowels, it makes up for in character. It's a bigger, fatter and hotter star than the Sun, and suffers from a severe case of heartburn as the contents of its stomach frequently bubble to the surface.

Technically, Przybylski's is an 'Ap' type star. The 'A' means it is hotter than the Sun, is blue in colour and has a surface temperature of around 10,000 degrees Celsius, and the 'p' stands for 'peculiar'. Yes, you heard me correctly. Peculiar is an official astrophysics word.

What is so peculiar about Przybylski's star is its chemical make-up.

We can tell exactly what chemical elements are present in a star by spreading out its light into a rainbow. By studying which colours have been nibbled out by the atoms in the star's cooler outer atmosphere, we can tell exactly how much of each chemical is present. This gives us an accurate idea of the composition of all stars – and we see strong patterns in what we observe.

Yellow/orange stars like our Sun are made up of 75 per cent hydrogen and 25 per cent helium, with other elements present in only tiny amounts (less than 0.1 per cent, in the case of the Sun). That's because the relatively low temperatures (only 15 million degrees) in the stars' engines allow for only a basic hydrogen-to-helium nuclear burning process.

This is in contrast to hotter, bluer stars, which are buzzing cauldrons of larger chemical elements. The higher temperatures of these stars trigger advanced nuclear fusion, which is able to generate enormous atoms from building blocks like carbon, nitrogen and oxygen.

The composition of Przybylski's star is peculiar indeed. It has far less iron and nickel than other stars of a similar temperature and far more of the rare heavy elements, such as strontium, neptunium, plutonium, curium and einsteinium. These rare atoms are radioactive, making Przybylski's star a fairly hair-raising environment.

The really weird thing about finding these radioactive elements in the star is that many such chemicals are short-lived. If you put 100 strontium atoms in a box and close the lid, then open it again after twenty-eight years, only half (fifty) will remain. The rest of the strontium atoms will have broken down or 'decayed' into other chemicals. That's what we call 'radioactive half-life': the average time it takes for half of the atoms to decay. For curium, the half-life is only 163 days.

Given that Przybylski's star is millions of years old, how did all the strontium and curium get (and remain) there?

This, my friends, remains a mystery.

Two theories have been put forward. The first is that Przybylski's star contains some as-yet-undiscovered heavy radioactive elements that are slowly decaying into elements like strontium and curium. That would be a highly significant discovery that could expand our knowledge of chemistry and the formation of chemicals in the universe.

The second idea is even more wacky: perhaps aliens poured these short-lived radioactive elements into the star just to draw our attention to it – a sort of 'signpost' for other intelligent life in the universe. Quite what they expect us to do next is a matter for us to ponder.

Another peculiarity is how the heavy radioactive elements get onto the outside of the star. Gravity causes the heaviest elements to settle to the middle – the nickel-and-iron core of Earth is a case in point. Given that strontium is eighty-seven times heavier than hydrogen, why does this star have so much radioactive crap floating about on its surface?

It's probably just indigestion.

Przybylski's is a rapidly oscillating Ap star, which means it 'breathes' in and out every 12.15 minutes. Although we don't yet fully understand how the pulsations work, it seems they are driven by the strong magnetic fields of the star. Pulsations serve to dredge up hot material – and with it, heavy elements – from inside the star and deposit it in the outer surface layer.

Either that, or extraterrestrial dump trucks are shipping in radioactive elements from some distant planet and sprinkling it onto the surface in an attempt to make friends.

Either way – don't you just love science?

The star with a diamond planet

Some stars are unusual, but not many qualify as unique. The quirky pair of PSR J1719-1438 and its 'diamond planet' are two such precious stones in a universe of pebbles.

A tiny, compact star that is smaller than the Earth, PSR J1719-1438 is a pulsar that spins on its axis once every 5.8 milliseconds. Imagine that: an object weighing the same as half a million Earths that spins on its head 200 times every second. A body roughly 19 kilometres wide that performs rotations ten times faster than your washing machine's spin cycle.

Studies of this pulsar have found that it contains a lot of carbon and oxygen – much more than the Sun, which is composed mostly of hydrogen and helium. Further observations have revealed that it is in an orbit with another body that has a mass equivalent to about 300 Earths. The pulsar and its companion orbit one another every two hours and ten minutes. We can't see the smaller body, but we can measure the gravitational disturbance it causes to the pulsar.

The secondary body was likely derived from a white dwarf star that probably lost 99 per cent of its outer gassy material as it was sucked onto the pulsar over the past few million years. As the pulsar gained weight, it began spinning even faster and the white dwarf shrank to its current 'bloated Jupiter' status. Now a dense remnant of its former stellar glory, the gargantuan planet is twenty-three times denser than water (that's twice as dense as lead). Since the star contains a lot of carbon, astronomers conjectured that the planet too (having been born from the same interstellar cloud of raw materials) would contain plenty of the element. With an extremely hard carbon core that is likely arranged in a solid lattice structure, the satellite was dubbed the diamond planet.

The world's media went wild at this story of a gazillion-carat star. A diamond orbiting a star!

Recent observations hint that we might want to be cautious with such cute monikers. New measurements suggest that the pulsar might not have as much carbon as initially thought, which puts a damper on the likelihood of it being a gleaming jewel in space. One researcher argued that the planet might not even be made of atoms at all but of strange subatomic particles called quarks.

Until we travel across the open plains of space and sample this enigmatic world, we may never truly know what the alien planet looks like. But whether or not the centre of it is made of solid crystallised carbon, the thought of a diamond planet moves us to imagine the rare and unworldly scenes that will await us when we finally break the shackles of our solar system.

7

Variables

Ancient folks, without the benefits of telescopes or modern light measurement techniques, assumed that the stars were permanent and immovable. Common sense dictated that since we don't notice any changes to the stars on a nightly basis, they must have been the same forever. Right?

Wrong.

The trouble with common sense is that it doesn't tell the whole story. Sometimes a comet would appear and track steadily across the sky for a few weeks, brightening and forming a beautiful tail before disappearing from view. Bursts of shooting stars would streak through the darkness and disrupt the nightly scene of serene perfection. Dynamic events such as these were considered scary warnings from angry gods that something terrible was going to happen.

It wasn't just moving objects that mucked up our ancestors' model of the universe. Every few years would see the coming of a new star – a nova, as the Romans liked to call them – that brightened and then faded away. These variable stars were also considered unlucky, because they were not understood as they are now.

All stars change in brightness to some degree. Our Sun hosts sunspots, which are darker than the regular surface of our star. They grow in number and size and then die off as the tides of the Sun's magnetic field wrestle with the convection of hot gas from below the surface. Like many stars, the Sun also experiences frequent burps, called solar flares, which increase its light output and spew high-energy particles and radiation into space and even into the path of the Earth. The Sun is also slightly wobbly on an eleven-year cycle, but its variations are very small – 0.1 per cent, to be precise (that's 1 part in 1000).

Stars can be variable for a few reasons. One is that pulsations are easily set up in their plasma, as the gas (which is effectively a fluid) tries to gain a balance between the hot air rising from the centre and the gravitational force pulling the star together. This tug-of-war sets up an unstoppable vibration, like the ringing of a bell, that makes the star brighten and fade like clockwork. Others fluctuate in brightness because they like to play hide-and-seek with their neighbours.

So who are some of these unlucky stars, these ancient trouble-makers of the cosmos?

The symbiotic stars

In the constellation of Cetus lies a glorious star called Mira. She changes dramatically in her brightness, with a uniformity and regularity that seem almost unnaturally regimented. Her variability has been known since at least the 16th century, when astronomers began documenting her comings and goings from naked-eye visibility every 332 days. Now you see her, now you don't.

Without an understanding of how stars shone or how they lived in pairs or groups, there was no way for astronomers in the Middle Ages to comprehend how a star might brighten and fade in this way. Even her name reflects her beauty and mystery, with 'mira' being the Latin word for amazing or astonishing.

With telescopes, the star was confirmed as a double. The larger member (Mira A) is a red giant star, a bloated remnant of a star like our Sun that has finished its steady hydrogen-burning phase and is now shedding its outer layers into space. The secondary star (Mira B) is a white dwarf, the compact remnant of a star's core. Hot and naked, dense, and with a gravitational pull beyond its diminutive stature.

This is no ordinary pair. Together, they form what is called a 'symbiotic' star system, the lean, muscly white dwarf pulling and tugging the gas from the large, ambling red giant, consuming its neighbour mercilessly.

Zooming in, you can see a powerful plume of gas spiralling in towards Mira B, forming a circular disc-like structure around the white dwarf star. A settled layer of gas like this is precisely the sort of environment from which planetary systems form. Only time will tell whether Mira B will spawn a new generation of worlds from the stolen stellar gas from its neighbour.

As well as this symbiotic relationship, Mira A has problems of its own. First, it is flying through space at 130 kilometres per second. Using ultraviolet cameras on a space telescope, we can see the effect of this supreme interstellar burnout unfolding. A large stream of hot gas is extruded from Mira A like the tail of a comet, stretching more than 100 billion kilometres into space.

Perhaps more alarmingly, Mira A is profoundly unstable. The red giant star undergoes transformative oscillations in which it expands and contracts so much that it occasionally adopts a 'potato' shape. As it does so, the star's brightness varies by a factor of fifteen times.

Although the Mira pair is not a perfect analogue of our Sun (after all, our star lives alone), it is true that the Sun will end its life as a red giant, probably huffing and puffing like Mira A. For all the differences, the famous variable star really serves to remind us of the future of our own solar system, when the outer layers of the Sun overtake the Earth and the other planets, transforming us into an interloper in an interstellar pair of bellows.

Makes you think, doesn't it?

The winking ghoul

The ancient Greeks believed that the stars were perfect, majestic, timeless and fixed in the sky. They were a permanent, predictable part of the firmament – the heavens – surrounding the Earth. This belief must have been comforting in a world of so much change and uncertainty – the faith that up in the heavens there was something we could rely on.

But just as reliable is that in a time of peace, a rebel will come in and break all the rules. Enter Algol.

Known in Arabic as *ra's al-ghūl*, this star represents the 'head of the ghoul' and is often known as the demon's star. In many mythologies it represents the head of the gorgon Medusa, a badass monster woman with snakes for hair who was said to be so terrifying that even giving her a quick side-eye would turn you to stone.

Is Algol the star really that terrifying?

Well, it does wink. Every two days, twenty hours and forty-nine minutes, like clockwork, it goes from being the second-brightest star in Perseus to 30 per cent fainter in a matter of hours. It quickly recovers, which to the ancients would only have proved its terrible supernatural powers. If you think about it, for people

2000–3000 years ago this would have seemed like a stupendous act of god-like magic.

We, of course, are armed with a greater superpower: science. With modern telescopes we can see exactly what's going on with remarkable clarity.

The demon star has now been uncovered by the powerful eyes of the CHARA (Center for High Angular Resolution Astronomy) telescope, an array of six optical telescopes, each with a mirror 1 metre across, that sit on a mountaintop in California. The zoom lens on this telescope is so powerful (200 times better than the Hubble Space Telescope) that it can actually take pictures of nearby stars that show them as spheres and not just points of light.

What the CHARA telescope found when it pointed at Algol was something remarkable that we rarely see in such an astounding level of detail.

CHARA has made a movie that shows a pair of stars caught in a cheeky foxtrot: in the centre, a hot blue star (Algol Aa1) that anchors the pair and appears steady and true; and in orbit, an orange star (Algol Aa2). Much further away is another orange star (Algol Ab) that does not contribute much to the party.

So close are the main pair of stars that the orange one is visibly stretched and distorted by the gravitational pull of the bigger blue star as it swings around on its wild ride. Like chafing on the inner thighs of a mechanical bull rider, the outer layers of the orange star are slowly being stripped by this constant pulling and pushing.

By chance, the spinning of these two stars happens to line up in almost exactly the direction that we view them from Earth. That means that every sixty-eight hours, the orange star passes almost directly in front of the bright blue star and for the next ten hours we enjoy the equivalent of a solar eclipse.

This is what causes the dimming of the demon star, not magic or the devilish powers of a snake-headed monster.

The midlife crisis

There are trillions upon trillions of stars that are visible in our universe with powerful telescopes. Would we even notice if one changed a little?

Most of the time, we probably wouldn't. We're not keeping tabs on each and every star, mainly due to the fact that most telescopes can only see a tiny portion of the sky at any one time. It's hard to monitor the sky in real time, even using wide-field cameras, since the amount of data gathered by these high-tech machines is too big for even the most powerful and capable supercomputers to process, transport and store.

It is often amateur astronomers in their backyards who first notice when something is amiss. That was the case for Sakurai's object, a faint 'new' star noticed by Japanese stargazer Yukio Sakurai in 1996. The matter was immediately referred to professional astronomers, who confirmed that the 'new' star had been missing from previous pictures of this part of the sky taken in the 1990s and 1930s. They confirmed Sakurai's object to be a nova.

With a bit more digging, scientists did find the star as a very faint blob on images from the 1970s, and then it was seen to brighten significantly during the mid-1990s, culminating in the nova event in 1996. Following this outburst, it faded again over the following two to three years as a pall of dark dust (as seen on infra-red images) surrounded it.

It turns out that Sakurai's object is going through a midlife crisis of epic proportions.

Have you ever seen a chef pour alcohol into a hot pan and engulf half the kitchen in flames? This is called a flambé, the French word for 'flamed'. We have the same thing in astronomy, but we call it a helium flash. That's what happened to Sakurai's object.

When a star has been working steadily in the same job (burning hydrogen and helium into carbon and oxygen) for billions of years and gradually inches towards the end of its career, it understandably starts to lose energy, to run out of steam. You could call it burnout, where the energy and the fire are no longer there. With fuel levels dangerously low, the engine of the star is running on empty. Technically, it is a red giant turning into a white dwarf. It has had its time burning hydrogen, and has also nearly finished helium burning. With little left to burn in its inert carbon and oxygen core, a shell of hydrogen surrounding the core burns feebly. Once enough helium has been created, this too is ignited. Woah. A colossal flash of heat and energy is released in this dramatic event – a helium flash. The star is given a second wind.

This is what Yukio Sakurai witnessed in 1996: a cosmic flambé, sending an ageing star towards early retirement as it fulfilled its destiny of going out with a bang rather than a whimper.

Space rulers

Imagine you're relaxing on the couch, a warm cup of tea nestled in your grateful palms. Your favourite book is just getting to the best bit. You close your eyes and take a deep breath that relaxes your whole body. Everything is good in the 'hood.

Suddenly, you hear a set of lightning-fast footfalls running down the hall. Great, the kids.

Next, a slow, lumbering footfall. That must be the partner.

Even with our eyes closed, our senses are familiar with the fact that smaller people have a faster cadence when they walk or run than people of a lankier ilk.

Cepheid variable stars are stars whose cadence reveals their size. They are the heartbeats of the sky – millions of hardworking metronomes beating their own rhythm. These stars have just the right temperature – not too hot and not too cold – to have zones inside their bodies where some helium atoms are partly broken up and others are completely smashed to pieces. When the atoms in the gas break up, they cool and collapse towards the centre of the star. As the gas falls, it compresses, becomes opaque again and absorbs heat, causing it to expand and once again become transparent. This sets up a cycle whereby the star 'breathes' in and out. As it does so, it changes its luminosity – the amount of heat and light it produces – on a regular schedule.

The variability of these stars is not just a gripping sideshow – it has been instrumental in revealing to us the true size and scale of our universe.

In 1908, astronomer Henrietta Swan Leavitt discovered a relationship between the size of a Cepheid variable star and how fast it breathes. The bigger and brighter the star, the longer it takes to take one breath. By measuring the time it took for one breath, Leavitt was able to calculate the intrinsic brightness of the star. This immediately gave us a tool to figure out how far away such stars truly are.

Given that Cepheid variable stars are so big and bright that they are visible through telescopes across vast distances of space, we have been able to prove that the so-called Andromeda nebula is actually a distant galaxy more than 2 million light years away. In the early 20th century, most astronomers thought this fuzzy, irregular cloud in the sky, and others like it, were simply part of our Galaxy, the Milky Way. The discovery of the true distance to other galaxies transformed our understanding of the universe.

A co-dependent relationship

We've all witnessed a co-dependent relationship. These unhealthy matches arise when one member of a partnership sacrifices all to attend to the whims of the other, putting the partner's happiness above that of anyone else, including their own. This often leads to friction and outbursts.

Stars, too, suffer from close relationships with destructive companions. A case in point is T Pyxidis, a sun-like star in the constellation of Pyxis (the mariner's compass). It is locked in a tight orbit with a heavy, dominant yet surprisingly small star, a white dwarf.

Since the sun-like star is less massive than its domineering partner, it gets slowly consumed by the tremendous gravitational forces emanating from the heavier companion. As the gas accumulates on the white dwarf it slowly builds up and acts as a blanket, increasing the temperature on the bottom of this layer of (mostly) hydrogen gas. Once the temperature of the consumed gas reaches about 10 million degrees Celsius, it ignites with a spectacular flash. It's like a new star has sparked into life! Nuclear fusion begins and the star becomes a nova, with a shell of hot gas ploughing a shock wave into space.

Over the past 150 years T Pyxidis has erupted six times, on each occasion becoming up to 2500 times brighter than it normally is.

There is evidence to suggest that these explosions fail to shed all the weight gained by the white dwarf, meaning that the central star is gradually growing in stature. This is a problem. Stars that eat gas from their neighbours can only 'nova' so far. A white dwarf that tips the scales over the magic number of 1.4 times the mass of our Sun immediately collapses under this load and destroys itself and any poor soul who gets in the way – that means you, sun-like partner – in a cataclysmic supernova explosion.

Will this happen to T Pyxidis?

Almost certainly. It may take 10 million years, or it could happen sooner if the eating competition continues apace.

Whenever it blows up, hopefully someone, somewhere in the universe is around to take in the spectacular light show.

8

Giants

Look at your hand. See the veins on the back?

Five litres of blood is flowing through your body right now. This life-giving liquid is tinted red by the iron that exists within every red blood cell, enabling the transport of oxygen around your body and fuelling its every movement, breath and function.

Iron is the most common element on Earth, by mass. It makes up the core of our planet, and its swirling motion as the Earth spins on its axis generates our planet's magnetic field. To this, too, we owe our lives.

The magnetic field of the Earth isn't just handy for reading a map with a compass. It is our last line of defence – an invisible shield that deflects billions of tiny but deadly particles from the Sun and from outer space that would otherwise bombard our planet and eventually kill us all.

Every atom of iron on planet Earth was made inside a star. That goes for every atom of iron in your blood, too. The vast majority of this iron wasn't synthesised by a boring middle-of-the-roader like our Sun, but by a supergiant star bigger than ten times the mass of our Sun.

These stars were born to rule. They create copious amounts of heavy elements and metals before leaking them back into the cosmos from their decaying bodies after they die. Their end is usually a great flash of glory called a supernova – a catastrophic explosion that acts like a thermonuclear bomb and synthesises even bigger and more complex atoms, such as gold and uranium and all the weird ones you've never heard of.

From these incredible exploding stars, our Sun and solar system were born. From the ashes of a long-dead star came our entire world.

Next time you look at your veins, or use a compass, or *don't* die in a shower of lethal cosmic radiation, be grateful to the unnamed relic of a cosmic behemoth who gave its life that we might live. Our ancestor star, our gentle giant of the skies.

Top Dog

The sizes and scales that we encounter throughout the universe are mind-boggling.

Take the Sun, for example. You could fit 1.3 million Earths inside the Sun if you had a big enough forklift. Some sunspots are bigger than planet Earth. Our star seems huge, huh?

Pfft. That's nothing. There is a star out there in the inky darkness of space that is so big you could fit 300 million Suns inside. If this colossal star were a removal truck, our star would be the size of an ant.

This celestial giant is VY Canis Majoris, a red hypergiant star in the constellation of Canis Majoris, the big dog. Let's call her Top Dog. She weighs as much as about twenty Suns and is so large that if you plonked her down in our solar system her gaseous bulk would extend from the Sun to the orbit of Saturn and we would all be engulfed in flames.

Top Dog is not just the biggest by size: she's also one of the most luminous stars in our Galaxy. She spews out 270,000 times more heat and light than the Sun yet is just over half the Sun's temperature. That's the power of being so large.

Why is she so big? She probably originated as a white-hot O-type star – that's the biggest and hottest type of star there is. At the time of her birth, she probably weighed as much as forty Suns. Her vast bulk came from the environment in which she grew – a giant cloud of molecular gas gravitating inside a large stellar nursery alongside hundreds of massive stars just like her.

O-type stars have cores that are tremendously hot from being squished and squashed together in one place. The heat sparks not only hydrogen fusion but also other processes like the so-called CNO cycle, which generates heavier elements like carbon, nitrogen and oxygen. These processes give off far more heat than the creation of helium in our Sun.

Like all dogs, Top Dog struggles to lose heat by radiating it from her body, so she resorts to panting. As convection set up in the core, radial pulsations began that caused her to swell up and lose mass into the interstellar environment. To this day we can literally *see* the oscillation of gas throughout the star and beyond, into her outer atmosphere.

Clumps of gas called masers (microwave lasers), made up of molecules such as water, silicon monoxide and hydroxyl, are lit up and shine with bright radio emission. We follow them using radio telescopes as they rise and fall in the star's atmosphere like helium balloons floating in the breeze.

Top Dog is the Drew Barrymore of the star world. Her success at a young age meant that she had to grow up very quickly. At only 1 million years old, she has lived less than a thousandth of the Sun's current lifetime but is already entering her twilight years. As her hydrogen and helium diminish and her light frame floats away, all that will be left is a dangerously heavy and shrinking core.

Without ongoing burning to keep the star looking youthful, her core contracts and a dangerous game begins. When her heart gives out and she collapses once and for all, Top Dog will meet a dramatic and final end as an exploding supernova.

Betelgeuse

What's big and orange and full of hot air?

I'm not thinking of a global political figure. Rather, I'm referring to Betelgeuse, the regal orange jewel in the famous constellation of Orion, the hunter.

Orion's distinctive shape has been recognised for tens of thousands of years in the folklore of peoples around the world. The figure of a hunter holds aloft a club and shield, with a sword hanging from his belt. Using the ancient Greek form, Orion's right shoulder is a bright orange star called Betelgeuse.

A red supergiant with around 8 million years of experience under his belt, Betelgeuse has led a similar life to that of VY Canis Majoris. Born with great riches as a supercharged hot blue star, he squandered this brilliant start in life, powering quickly through his fuel supplies before swelling up to the bloated proportions he has today.

Put Betelgeuse in the place of our Sun and his bulk would engulf the orbits of Mercury, Venus, Earth, Mars and Jupiter – not to mention the asteroid belt, with gas that crackles with heat at 3200 degrees Celsius.

Betelgeuse is so large that we can actually take a picture of his body and monitor the subtle changes in the shape and size of gas exhaled into his enormous atmosphere. Telescopes such as the Hubble Space Telescope and the Atacama Large Millimeter Array, in Chile, have made direct images of Betelgeuse and measured his size directly – something that has only been achieved for a small number of relatively close, large-diameter stars.

Astronomers keeping an eye on this orange monster have also detected changes to his size (he seems to be shrinking), although these changes could simply be attributed to the pulsations he frequently undergoes. Scientists have also identified hotspots on his surface, which may represent the emergence of enormous convection cells.

For all his bluster, Betelgeuse appears to be fading quite dramatically. At the start of 2020 he was the coolest and faintest he had been in twenty-five years and showed no sign of recovery. I've been looking at Orion since I was knee-high to a grasshopper, and even looking at Betelgeuse from my bedroom window now, it's very obvious to me that he is about two and a half times fainter than normal. Whether this is due to his natural cycles of pulsation or whether he is genuinely flagging, we cannot be certain.

Many have been asking: is this it? Has Betelgeuse's heart finally filled with iron and is it time for him to explode?

Reasonable expectations are for Betelgeuse to explode as a supernova sometime in the next 100,000 years. When he does, things will get pretty spectacular. His glowing gas will expand into space and the remnants will outshine our full moon. Betelgeuse will remind us of his presence as a shining light visible during the day for several weeks after the funeral.

Even though the odds of witnessing this in my lifetime might only be around 0.04 per cent (assuming I live to eighty years old), I'm kind of hoping I do get to see the demise of this puffed-up orange beast. If I were to witness a supernova that remained as bright as the half-moon for a full three months, it would be a memorable and fitting end to an old friend and a giant of our skies.

Meet big Ray

Want to meet the brightest known star in the universe?

Before you get too excited, a few caveats. First, we can't see individual stars beyond the nearest handful of galaxies, so there might be meatier specimens further away. Second, there are an estimated 70 sextillion (70,000,000,000,000,000,000,000) stars spread out over 2 trillion (2,000,000,000,000) galaxies in all the space we can possibly see (the 'observable universe'). Beyond that we will never know how many stars there are, nor how bright they might be, since the relentless expansion of the universe means that the light from any star that shines from beyond that point will never reach our part of the universe.

Putting all of that aside, do you want to meet the biggest and brightest star we know?

Meet R136a1 – let's just call him Ray for short – who tips the scales at 315 times the mass of our Sun and outshines her by 8.7 million times.

Ray is a Wolf-Rayet star, a special type of supergiant who continually pushes scorching gas from his body into space at frightening speeds in excess of 1000 kilometres per second. Over hundreds of thousands of years, this has led to the accumulation of a sweltering bubble of gas around him that glows in the monochrome hues of so-hot-they're-smashed-up hydrogen, nitrogen and helium atoms.

With a group of heavyweight siblings, Ray lives in the nearby galaxy called the Large Magellanic Cloud, some 160,000 light years away. Together, this clan of young and ambitious athletes have created their own ecosystem, the Tarantula nebula, whose swirling gas tendrils glow and sing in many colours across the cosmos.

The colours of stars like this are fascinating. So intense is that furnace of fusion in Ray's belly that his surface temperature eclipses our Sun's by a factor of ten. His colour can best be described as ultraviolet.

You heard that right.

As you may recall, all stars shine in a broad range of colours (rainbows, anyone?) but cooler stars 'peak' in the red end of the spectrum and hot stars emit more in the blue colour palette. That's down to how energy is transmitted when particles move around in a hot substance. Think about that difference between an orange candle flame and a blue gas flame.

Ray's gases are so hot that his light peak is beyond blue – it's invisible to the human eye! Luckily, since stars emit a range of colours (peaked around the ultraviolet, in his case), some visible light sneaks out and the colour we perceive from him is blue.

The sting in the tail? With Ray living so far from Earth, we can't be certain of some of his physical attributes. His temperature is estimated from the complex mix of colours we perceive from his light, and his size cannot be determined directly by taking a photograph – he just looks like a point of light.

As we learn more about our universe and discover greater things, it is surely only a matter of time before Ray loses his heavyweight title to another star.

The Big Friendly Red Supergiant

One of my favourite books when I was a child was Roald Dahl's *The BFG*, a heartwarming tale of a 28-foot-tall giant who meets a little girl called Sophie. At first she is afraid the giant will eat her, but he reassures her that he is the Big Friendly Giant and nothing bad can happen to her. Since Sophie has no parents, he takes her in and feeds her, and they embark on all sorts of exciting adventures together. The BFG protects her from being eaten by other, less friendly giants, and in return Sophie teaches him to read and write. He builds a wonderful castle where he lives and a cottage for Sophie next door.

I recalled this story as I studied one of the most stupendous giant stars in our Galaxy, VV Cephei. This red supergiant is like the BFG in many ways, since it is very big (around 1 billion kilometres across) and generally very friendly. It is also visible at night-time to the unaided eye.

Annie Cannon from Harvard College Observatory noticed in the 1930s that the light from VV Cephei was varying in a regular way, and she and her colleagues soon figured out that this was a binary system. The light patterns measured revealed that a smaller blue star was orbiting VV Cephei A, eclipsing it as it orbited in a period of over twenty years.

We have found our Sophie! Just like the BFG, our Big Friendly Red Supergiant (BFRS) shelters Sophie and looks after her. He keeps her close. If the BFRS replaced our Sun, his surface would reach almost to the orbit of Saturn, and Sophie would orbit somewhere between the orbits of Uranus and Neptune.

Sophie's proximity to the BFRS and his low-density gas mean that he literally feeds her. Now Sophie is no slouch – she is a bright yet compact B-type star with her own supply of fuel. But as she orbits the BFRS, gravity does its work and she slowly consumes gas from his outer layers that curls towards her and arcs around into a gas disc, which she keeps for later like oats in a horse's nosebag.

We can't see Sophie's disc of hot gas directly because she is too far away, but its bright light is visible when we break her light into a rainbow and see the distinctive colours of searing-hot gas surrounding the star, including heavy metals such as iron, copper and nickel.

Every twenty years, for eighteen months, the BFRS eclipses Sophie, hiding her completely. One can only imagine this is to protect her from being eaten by those other, not-so-friendly red supergiants out there in the Galaxy.

Now that would be scary.

The Pistol Star

Have you ever done something really cool, like a trick shot in basketball, but there is no-one there to see it?

Spare a thought for the Pistol Star, which has spent its whole life doing truly phenomenal things but is completely invisible to the only known spectators in the universe. (That's us, by the way.)

Shining with a brightness of 10 million Suns, the Pistol Star is a massive unstable blue hypergiant that periodically undergoes outbursts and eruptions that shed weight faster than when you have a bout of gastro. It had one massive ejection around 4000 years ago when it spewed around 19 million trillion trillion kilograms of gas into space. (It was actually around 30,000 years ago, but the light has taken 26,000 years to reach us, so we're a little behind.) Not only does it experience explosive events like this, but it is also continually undergoing extreme mass loss from its stellar wind. That's the constant torrent of particles streaming from the star – and the Pistol Star's wind is *ten billion* times stronger than our Sun's.

All of this material that the star has discarded now hangs around as the glowing Pistol nebula.

That's all very impressive, but until the early 1990s we didn't know any of it existed. Despite being one of the most luminous stars in the Galaxy, the Pistol Star is completely invisible. We can't see it at all with an optical telescope – we can only see it by zooming in with an infra-red camera on the Hubble Space Telescope.

By rights, the Pistol Star should be easily distinguishable to our eyes in the constellation of Sagittarius. But there is something in the way. That something is our Galaxy's thick layers of black dust.

The Milky Way is a spiral galaxy containing around 300 billion stars that are arranged into several spiral arms containing stars and gas. We live in the Perseus spiral arm. The arms are concentrated around a dense mass of stars, gas and molecular clouds that make up the central zone of our Galaxy.

The spiral arms of the Milky Way are visible as a brilliant band of light, made up of billions of stars, that arcs across our night skies. The core of the Galaxy lies in the direction of the Sagittarius constellation, where dark lanes of dust abound and with a quick scan you notice far fewer stars. If you know the constellation of the Dark Emu, this is where you'll find it. The darkness you see in these parts of the sky is not because there are fewer stars – it's simply because they are hiding behind dark molecular clouds.

The Pistol Star was a gem waiting to be discovered. Now, aided by the scientific dexterity of our minds, and technology that enables us to see what is invisible to our senses, this cosmic giant is within our grasp.

9

Self-destructors

Some of the astronomical phenomena I have witnessed are etched in my memory forever.

The aurora dancing across the sky in bold curtains of green and purple light. The stillness of the air and the cool touch of dusk during the total solar eclipse of 1999. A shooting-star 'fireball' flashing like a brilliant green laser light through the darkness. The two great comets of my childhood, the likes of which have not since been seen.

Sadly, there are some things on my astronomical bucket list that I am yet to witness. A supernova is one such rare jewel. Neither I nor most other human beings will ever be lucky enough to see one.

A supernova looks like a dazzling star in the sky. But the treasure lies in the understanding of what it represents: the cataclysmic destruction of a star that has lived for millions of years and, with it, a shower of unimaginable heat, radiation and light outshining its host galaxy for days or weeks.

During a supernova explosion the entire contents of a star empty into space, enriching it with chemicals that may one day make up a new star, a planet, or even a sentient being like you or me. In many cases the event could also signal the destruction of a solar system of planets that have orbited their parent star faithfully for their entire lives.

The sight of a supernova is rare. In the Milky Way, with an estimated 200 billion stars, we expect to see an average of three supernovae every century. That estimate is based on the number of supernovae recorded over the past few hundred years. What's more, our Galaxy is so full of dark molecular clouds that more than half of these supernovae will be completely hidden from view.

The first supernova for which we have fairly unambiguous written records happened in the year 185 CE, when a Chinese historical account described a new star shining for between one and two years in the constellation of Nanmen (Centaurus). Three supernovae were then recorded in the mid to late 300s, followed by a drought until 1006 CE, when the brightest stellar explosion in recorded human history went off.

The light from supernova 1006 was described in Chinese records as 'huge ... like a golden disc. Its appearance was like the half Moon and it had pointed rays.' Another record said that the new star was 'so brilliant that you could see things clearly at night [by its light]'. The supernova was visible well into 1009, three years after the original event.

Imagine witnessing something like that! A new star that shines bright like the Moon every night for months on end.

Only forty-eight years later, our planet was treated to another bright supernova, SN 1054, which left a spectacular remnant of glowing gas called the Crab nebula that still hangs in space like a freeze-frame from an explosion in a movie (google it for awesome pictures). Deep in the heart of this nebula sits a neutron star, the last, tiny part that is left of the once-glorious star. After a faint-ish supernova 100 years later, we had to wait 400 years before another spectacular supernova came along, in 1572. The detonation of this star shone brighter than the planet Venus and was visible during the daytime. Imagine how excited Danish astronomer Tycho Brahe must have been when he saw it and then worked for weeks to carefully record its radiance as it ramped up and then slowly faded from view.

The last supernova to be seen in our Galaxy was SN 1604. Chinese, Korean and European astronomers made hundreds of measurements of the brightness of this explosion and its aftermath as it glowed with a light visible to the naked eye for more than a year. Even now, through a decent telescope, we can see the shell of radiating gas expanding into space where a once-mighty star used to be.

More than 400 years have passed since then. The next supernova is long overdue – but which giant star will do the honours? Will it be Betelgeuse, Rigel, Spica or Eta Carina?

Let's hope it's not too close to Earth. Tin hats on, everyone!

Hello, star

Despite being incredibly rare, supernovae have to happen.

The one we're about to meet picked a Monday morning, 23 February 1987. Bon Jovi's 'Livin' on a Prayer' was number one on the charts and I was seven years old. At about 7.30 a.m. I'd just woken up at home in Wethersfield, England and was eating breakfast (Marmite on toast) and getting ready for school, oblivious to the cataclysmic astronomical events that were about to unfold.

In the Large Magellanic Cloud, 168,000 light years away, a blue supergiant star, SK-69 202, had its own plans for this Monday morning. Triggered by a series of events that we still don't fully understand, its core was about to collapse – and quickly.

Five minutes later, inside the Kamiokande-II neutrino detector deep within a disused zinc mine in Japan, planet Earth received the first hints of what was happening in our neighbouring galaxy.

Inside the mineshaft, a shower of invisible particles called neutrinos – tiny subatomic particles with almost zero mass that are generated when particles collide at high speeds – streamed into an enormous underground tank of water, some of them crashing into the nuclei of the hydrogen and oxygen atoms and triggering flashes in the detector's cameras.

Yep, this was highly unusual.

Back in the Large Magellanic Cloud, SK-69 202 was in serious trouble. The neutrino burst detected in Japan was a sign of its catastrophic implosion. When the star was no longer able to generate energy by nuclear fusion, its weight became too great and the core suddenly collapsed. Unstoppable, like an avalanche, it reached speeds of 750,000 kilometres per hour.

In less than a second, the star's engine reached a temperature of 100 billion degrees. Electrons crushed together with protons and combined to form neutrons, the neutrally charged particles that live in the middle of atoms. This produced the outburst of neutrinos that streamed across the universe in all directions and was detected 168,000 years later on a Monday afternoon (local time) at Kamiokande-II.

Neutrons are tough buggers. They don't take kindly to being squashed together, and when they are, they form a defence stronger than the New Zealand All Blacks back line. In this neutron-only environment the brakes slammed on and the collapse of the star was immediately halted. The true moment of supernova happened when the unimaginable momentum of a trillion trillion tonnes of gas rebounded off the brick wall of neutrons. The shock wave ripped the star apart in the blink of an eye, spilling anything and everything into space. Neutrons were literally forced together with heavy elements like iron and nickel and transformed into even bigger atoms like gold and uranium.

Around 10.30 on that Monday morning, the school bell rang for morning break and I dashed outside to play football with my friends. While I was recreating goals from the 1986 World Cup, on the other side of the world, light from the supernova explosion was reaching Earth. Within hours, observers in New Zealand and Chile had noticed the new star in the Large Magellanic Cloud. Given the moniker Supernova 1987a, it was the first naked-eye supernova seen since 1604 and was visible for several months.

Thanks to modern telescopes, we learned a lot from Supernova 1987a. The detection of neutrinos showed categorically that type II (type two) supernovae are triggered by the catastrophic collapse of a star's core. We watched as the shock wave of a supernova blasted into space, illuminating beautiful rings of gas that surrounded the progenitor star before it exploded. And we measured in great detail the chemical evolution of the star as 20,000 Earth masses of iron was thrown from the explosion and a dusting of molecules like carbon monoxide and silicates emerged from the ashes.

More than thirty years have passed since that event, but we are still uncovering secrets from Supernova 1987a. For me, most exciting of all is the chance (once the clouds have dispersed) to find out whether a neutron star or a black hole was left behind.

The big one

On 13 May 1972, Charles Kowal, an astronomer at the Palomar Observatory in San Diego, California, noticed that an extremely bright new star had appeared in NGC 5253, a small irregularly shaped galaxy almost 11 million light years from Earth. He quickly alerted his astronomical colleagues that something unusual was underway, and observers around the globe set to work monitoring what could only be explained as one of the most powerful supernova explosions ever seen. Optical, infra-red, even X-ray and gamma-ray telescopes were deployed to probe exactly what radiation might be coming from this exploding star in the depths of space.

The 'light curve' of the stupendous star – a graph of how quickly it brightened and then dimmed – showed a very familiar pattern, obvious to astronomers as the fingerprint of a type Ia (one-a) supernova.

Type Ia supernovae occur when a white dwarf star lives in a binary system and starts to consume the gas from its companion. The unwilling donor of the gas could be a giant star, a sun-like victim or another white dwarf. As this stellar material is piled onto the tiny white dwarf's frame (these stars are only the size of Earth), the white dwarf starts to feel the strain. Its temperature and density increase, its knees begin to shake under the weight, and eventually a lightning flash of carbon and oxygen fusion is triggered.

Normal stars are able to regulate their temperature by expanding when they heat up, but white dwarfs are different. Their atoms are squashed together into an unimaginably dense type of material called 'degenerate matter', in which the usual laws of thermal regulation do not apply. When the flash of carbon and oxygen fusion begins in the white dwarf, it heats the star to more than 1 billion degrees Celsius in just a few seconds. The star's internal forces can't contain this energy and the star detonates. A supersonic shock wave rips through the poor star's body at a speed in excess of 10,000 kilometres per second and it explodes like a bomb. Its guts and still-burning gases spill out into the universe. Freedom!

Type Ia supernovae are 5 billion times brighter than the Sun at their point of maximum brightness, which occurs a few days after the explosion. After that, the remnants of the star fade gradually over a few months in a predictable way.

In 1972, supernovae were less well understood than they are now. Astronomers watching Supernova 1973e (as it was now called) found that the rate at which it was dimming agreed with a theoretical model that had recently been put forward about how the gas ejected by the supernova keeps glowing after the explosion is long gone. In this model, the explosion triggers nuclear fusion of heavy elements in the star, producing large quantities of a radioactive form of nickel, the silver-white metal we use in stainless steel and batteries. With a half-life of six days, the nickel decays quickly into radioactive cobalt, which also decays over the next seventy-seven days. These radioactive processes provide the energy that lights up the supernova even weeks after it has exploded. The star that provided the energy has long gone, splattered across the universe, but its corpse shines on.

Careful observations of Supernova 1973e, and others since, have cemented our understanding of what powers type Ia supernovae. We now use them as 'standard candles' to measure cosmic distances. If we know how bright they *actually* are, and how bright they *appear*, we can easily calculate how far away their host galaxies lie.

This simple process has led to much of our understanding of the size and extent of the universe. It has uncovered the still-unsolved mystery of how and why that expansion of the universe is accelerating. Is the universe filled with a mysterious substance called dark energy, or are we missing something else? Astronomers are working night and day to find out.

So thanks, SN1972e, for shedding so much light on our universe. What an incredible achievement for an Earth-sized star lying 11 million light years away.

How peculiar

You can't enforce uniformity on people, and you certainly can't make exploding stars behave as you want them to.

For many years, astronomers wanted to believe that all type Ia supernovae behave in a completely standard way. Like clockwork, they explode, their remnants brighten and then they fade again. Scientists never questioned whether it was acceptable for honking-great stellar explosions to stray from the path of righteousness, or whether they ever would. Longing for uniformity in these 'standard candles' was a natural explanation, since the importance of type Ia supernovae in measuring the distance to faraway galaxies and understanding the evolution of the universe depended on their total predictability.

In 1991 that all changed. Within a year, two very weird and rebellious type Ia supernovae came along to break the assumption.

Supernova 1991T was a bit odd. It revved up much faster than an ordinary type Ia, was brighter at its peak, and had a different chemical mix. Astronomer Brian Schmidt called it 'the black sheep of the type Ia family'. The reasons for its oddness were not understood, but researchers guessed that the white dwarf star at the centre of the explosion had failed to detonate fully, or its detonation had been disrupted in some way, leading to the unusual progression of the supernova.

The same year, another so-called type Ia supernova dared to be different. Supernova 1991bg had a lower maximum brightness and faded faster than regular members of its clan. Astronomers waved this inconsistency away as 'Perhaps the star exploded funny' or 'Maybe two white dwarfs collided'.

A few more misfits were discovered in the ensuing years, and these quirky rebels became known as 'peculiar' type Ia supernovae, seemingly rebels without a cause (or an explanation). No scientist has successfully recreated this type of supernova in a computer model, so for now the cause of the rogue behaviours remains a mystery.

Perhaps the most peculiar type Ia supernova came along in May 2002. Extremely faint, and visible only through a very large telescope, Supernova 2002cx was discovered by astronomers at the Palomar Observatory. Its light curve was different from normal type Ia supernovae (and from the other oddballs I have mentioned) and its chemical mix was way off the charts, with unexpectedly small quantities of calcium, sulphur and silicon appearing as the entrails of the star's gas cooled in space. This supernova was dubbed especially 'astonishing' (in the words of one researcher) because of its relatively low peak brightness and the glacial rate of its expansion.

In 2002cx the torrent of high-velocity iron atoms that is usually the hallmark of a type Ia thermonuclear explosion was replaced by a whimper of a display, like sodden fireworks in a New Year's Eve downpour. A few weeks after the explosion, leftover gas hissed out at the relatively slow rate of 700 kilometres per second, the supernova equivalent of air leaking from a punctured tyre. Some of this gas seemed to be unburned, containing gaseous elements like oxygen that would usually have been converted to heavier elements in the explosion.

Could this mean that the detonation of this supernova had been halted or disrupted in some way, or that the runaway fire front had not reached all of the white dwarf? In other words, was 2002cx a failed type Ia supernova? Or, as one researcher has suggested, was there an element of core-collapse in this supernova and was it a type Ia/type II hybrid?

The simplest answer is the right answer: we simply don't know for sure. As our cameras get better and more supernovae come within our grasp, we will certainly discover more erupting stars that deviate from the norm. Hopefully, this will herald a new age where diversity in supernovae is seen as something to celebrate.

The gassiest baby in the universe

Stars have a lot of ways to expel gas. Like babies, they are constantly eating, burping, farting, puking or otherwise leaking gases from their bodies. That is part of their charm.

The winds from stars pour a constant stream of hydrogen, helium and high-energy particles into space from the day a star is born. Flares from dwarf stars throw concentrated burps of stellar material into the interstellar void. Giant stars puff up and pulsate, shedding their outer layers and forming stunning nebulae that decorate our universe like paintings on the walls of a fine art gallery. Highly evolved massive stars erupt and detonate, spewing trillions of trillions of tonnes of complex elements into space, enriching the chemical make-up of our universe and paving the way for the creation of life.

Eta Carina is a star whose unpredictable behaviour surpasses all of these expulsions. Meet the gassiest baby in the universe.

Actually a binary pair of stars, Eta Carina is dominated by an unusually bright luminous blue variable star, which is orbited by a bright massive blue star. We can't see the stars directly (i.e. take a photo) because they are surrounded by so much cloudy material – testament to thousands if not millions of years of the star's wailing tantrums.

The last time Eta Carina spat the dummy in a major way was 1837, during the so-called Great Eruption. Visibly, the main star transformed overnight from fairly bog-standard to one of the most brilliant stars in the sky. Its brightness wobbled a couple of times during the outburst, then the star faded slowly over the next couple of years. During its eruption, it threw 40,000,000,000,000,000,000,000,000,000 kilograms of gas into space. Over the coming years it slid completely from view as it was enshrouded in a dense fog of its own creation.

This is not the only temper tantrum Eta Carina has thrown. It erupted again in 1887, but since it sits in its own mess – a peanut-shaped cloak of thick smoke called the Homunculus nebula that envelops the star – this outburst was barely noticed. Forensic study of the shape and extent of the nebula surrounding Eta Carina suggests that similar outbursts probably occurred in the 13th century and the mid 16th century.

Over the past hundred years, Eta Carina has grown up somewhat. No major instances of spitting the dummy have been noticed, although the star fluctuates frequently between being bright on the one hand, and invisible on the other. Given that Eta Carina has been brightening steadily since the 1930s, is it perhaps building up a head of steam for the next outburst?

All we can do is keep our eyes on the skies.

The star who cried wolf

When I was a child I used to read Aesop's Fables, a collection of moral tales. One of my favourites was 'The Boy who Cried Wolf'.

A young shepherd is tending to his sheep on a hill when, bored, he decides to cry out 'Wolf! Wolf!' The villagers dash up the hill to help but when they arrive there is nothing there. Muttering their annoyance, they return to the village. The boy sniggers.

The next day, he calls again: 'Help! There's a wolf!' The people dash up once more, but when they arrive there is no wolf. The boy grins and watches them go grumbling back down the hill again.

A few days later, a wolf appears and chases the sheep. The boy cries 'Wolf! Wolf!' and the villagers ignore him. His flock is killed and the boy learns a valuable lesson: don't push your luck.

Supernova iPTF14hls could learn a lesson from this tale. It started fairly normally when, in 2014, a new, bright spot of light was found in an image of a galaxy lying half a billion light years away. The event seemed at first to be a fairly ordinary core-collapse supernova (if you can ever call the cataclysmic destruction of a star ordinary). But rather than cooling down and fading after about 100 days, as a well-behaved type II supernova should, it kept flaring and shining brightly for almost three years.

Cue much head-scratching from astronomers. They dug through the archives and looked at old photographs of this region of the sky, and found several images that on closer inspection showed that the star had 'previous' (as we say in London). It had pretended to go supernova before, like the boy who cried wolf.

In 1954, astronomers accidentally snapped this galaxy while the 'supernova' was erupting. By 1993, all signs of the explosion had gone. It appeared again in 2014, quite a feat for a star that, by all accounts, had been destroyed fifty years previously. It seemed that Supernova iPTF14hls was faking it.

How many times had it pulled this trick before? And what could account for the star's repeated, variable and lengthy supernova-like histrionics?

As with every astronomical mystery, several competing theories have been put forward, tested and discarded by scientists. That's how science works, and it's an interesting and creative process.

One theory being explored is that the repeated pulsations of the star are caused by antimatter created in its core. Antimatter sounds like science fiction but it is 100 per cent real – we can make it in particle accelerators. It's just like normal matter but has the opposite electrical charge, so an anti-proton has the same mass as a proton but is negatively charged, and an anti-electron has the same mass as an electron but is positively charged.

The insides of very massive stars are just like particle accelerators, with high-energy particles and gamma rays crashing into one another. From these collisions, it is thought that antimatter particles could be created. But making antimatter takes up a lot of energy, which is drained from the core of the star and causes it to shrink. As it does, it heats up and at a critical temperature the oxygen in the core ignites, causing an eruption of energy that looks like a supernova but does not destroy the star.

Sound plausible? It's a maybe, but the physics doesn't quite add up. The amount of energy predicted to be generated by this process in computer simulations is not enough to explain the 2014 outburst of our prima donna star.

Another theory is that the 'eruptions' of iPTF14hls are simply variations in the stellar wind, which blows a tremendous gale as a 'superwind' and sometimes whips up to become a tempestuous 'hyper-wind' that throws the gas equivalent of ten times the mass of our Sun into space every year. As this crashes into the gas that already surrounds the star (having been tossed there in the superwind phase), the gas is lit up like a gargantuan space light bulb, which is what we see as a flaring 'supernova'.

A definitive understanding of the behaviour of this fascinating star eludes us.

Should we forgive iPTF14hls for crying wolf? It never claimed to be a supernova, although it did dress as one. Maybe it's a wolf in supernova's clothing.

10

Runaways

I was sixteen when I saw my most memorable astronomical sight to date: the great comet of 1996.

It was first spotted by an amateur astronomer called Yuji Hyakutake (through a pair of binoculars!) at the end of January that year, and the astronomical grapevine began to buzz. But there were many unknown factors. Would the trajectory of this icy rock from the distant outer solar system bring it close enough to the Earth to make it visible to us mere mortals?

Comet Hyakutake didn't disappoint. By March it could be clearly seen without binoculars, and as it careered towards the Sun, our star heated the surface of the comet. The various chemical ices on the face of the rock, including carbon dioxide, ethane and methane, turned into gases. This caused an enormous glowing gas 'atmosphere', or coma, to form around Hyakutake's rocky surface and a tail to stream away from the Sun. The tail is not formed by the swift movement of the comet leaving the gases behind, as you might imagine. It is actually swept back by the solar wind – all those hot particles flying off the Sun as it burns hydrogen to helium in its core.

I will remember vividly and forever the moment I first saw it. It was late in the evening and the sky was clear, which is quite rare in England. I knew that the comet should be visible towards the north, so I walked down Meadside, the small street of council houses where I grew up, and crossed the road to stand in a small pool of relative darkness that was shielded from nearby streetlights. Glancing back up towards the sky above my house, I saw it. The comet was unmistakable against the familiar stars. I felt like a truck had hit me and knocked every bit of air from my lungs.

The comet was big – bigger than the full moon – and a strange, almost unnatural green colour. Quite unlike a star or planet, its head had a fuzzy outline and its tail was white, long and serpentine, and more perfect and beautiful than I ever could have imagined. It was an incredibly moving and emotional experience to finally witness something I had only read about in books – a great comet that comes without warning or invitation to put on a spectacular show.

People on planet Earth have long been awed by transient astronomical events like these. Seeing a new star, or a rock from the outer reaches of our solar system, whiz past at 150,000 kilometres per hour broke the spell of the apparently static universe that was presented by the serenity of a clear, dark sky. As we learn more about our cosmos, it is clear that not only is everything transient, not only are stars born and they eventually die, but also nothing in the universe sits still.

The Earth circles the Sun at an average speed of 100,000 kilometres per hour, completing one orbit every 365.25 days. The Sun cruises through our Galaxy at 828,000 kilometres per hour, completing one lap in around 250 million years. The Milky Way is like an interstellar merry-go-round, with every star and every cloud of gas gliding peacefully in circles. As the Sun and the other stars orbit our Galaxy, the relative positions of stars change and the constellations gently evolve and melt away.

But there's always one, isn't there. The screaming kid running through a restaurant. The driver on the wrong side of the freeway.

There are a few stellar scallywags who break our Galaxy's peace. Reckless rascals who career through the Milky Way with no care for the rules of the road or the trails of destruction they leave as they streak through our skies like the wandering comets in our solar system. Stars who break free from the bounds of our Galaxy and dare to venture where no star has gone before. Interstellar interlopers in our midst.

Ah yes, the runaways.

The Milky Way matriarch

Our first runaway star is a wise old traveller. Her official name is Kapteyn's star, named after the man who discovered her unusual properties. But we're not yet living in the dystopian novel *The Handmaid's Tale*, so let's give her a name of her own. Let's call her Barbara.

Despite being one of the closest stars to Earth, Barbara is her own woman. For starters, she is a master of hide-and-seek. She hides in plain sight, invisible to us but only 13 light years away. A fading subdwarf with less than 30 per cent of the Sun's size and mass is not easy to spot, even at such close proximity.

A woman never reveals her age, or so goes the old trope, but our Babs is estimated to have wandered our Galaxy for a knee-aching 11 billion years. We know her age because she's old school – it's in her DNA. Barbara shuns newfangled inventions like oxygen, carbon and silicon: her body is an almost perfect mix of the hydrogen and helium that was generated when the universe was small and hot just after the Big Bang. Back then our universe simply didn't have heavier elements. It would be a long time before the subsequent generations of stars burned their hydrogen and helium to make these elements and spill them into space to create new stars. It is because Barbara is made of this virgin gas that we know she's so old.

She might be old, but boy is she fast. Around 10,000 years ago she was one of the closest stars to Earth, only 7 light years away.

Since that interstellar fly-by, she has been rocketing away at 245 kilometres per second. In fact, her blistering speed is what got her noticed in the first place.

She doesn't glide through the disc of the Milky Way like the Sun and all the other stars. Her path is very different. Barbara is a wandering star on a retrograde orbit. By carefully measuring her speed and path through the sky, we can figure out where she's been and where she's headed.

The answer is impressive. It turns out that this unassuming matriarch of the Milky Way is actually from somewhere further afield. She's come from far away, outside the spiral arms of the Milky Way. An ancient invader, visiting our Galaxy's disc but not bonded to it.

And she's not the only one. Scientists watching the weird movements of nearby stars have found that Barbara is just one of a marauding group of grey nomads slowly touring our Galaxy on a circuitous path. Their speed, trajectory and chemical make-up point to a likely origin in the halo of our Galaxy – a massive sphere of ancient stars and star clusters that lives in the outer suburbs above and below the Milky Way's disc.

Some astronomers have suggested that Barbara and her travelling companions come from a prehistoric globular cluster called Omega Centauri within our Galaxy's halo, which could be the remnants of a long-dead dwarf galaxy that collided with the Milky Way in the distant past. This has recently been refuted by careful comparisons of the amounts of each chemical within the grey nomad group and the Omega Centauri cluster, which do not match.

Barbara is a free spirit, after all.

The jilted lover

We have met the star families – pairs or groups that bond through mutual gravitational attraction, unite in orbital matrimony and muddle along together in stable relationships for the remainder of their days.

As in life, though, relationships like these are not always plain sailing. It's not all country walks and white picket fences. Families can sometimes fracture and break down – and the families of stars are no different.

Let's say that a star who is already spoken for has his head turned by another. A mutual gravitational attraction is set up. What happens then? These situations rarely go well, and infidelity is a distinct possibility. The temptation is particularly acute for massive stars (more than eight times heavier than our Sun), because most of them live inside communities filled with temptation. They are born in gigantic molecular clouds where gravitational collapse is rife and conditions become ripe for the creation of thousands of stars, great and small.

The intensely cramped conditions within these star cities means that sometimes closeness is inevitable. In the course of their random wanderings, a married couple will often sidle up close to a single star. Attraction is not a choice where gravity is concerned, so it is not uncommon for a coupled star to take an interest in a bright, shiny young singleton. That's when the trouble begins.

Couples thrive on stability (gravitationally speaking), and proximity to another star, especially a big heavy one, tips the delicate orbital balance between the pair. The couple's graceful equilibrium is quickly distorted by the insatiable draw of the interloper, and the stars collapse into a gravity-bound three-way fight to the death. Only luck (and relative mass) will determine the outcome as they fall, whiz and bounce together into an unstoppable cosmic spin. In the melee, the biggest two stars win and pair up, and the other is ejected.

Our jilted lover flies helplessly away from the group, accelerated to more than 50 kilometres per second. The star itself is unchanged – only its speed and direction are altered as it rockets through space. As its powerful stellar wind of particles streams out in all directions, a bow shock forms ahead of it, like the wave pushed forward at the front of a boat. Behind it, a trail of tears is left in its wake.

This star is an outcast, bereft and wandering alone through the Galaxy. Robbed of its partner and companion by the cruel force of gravity, its only hope is that one day, another star will come wandering close and will find it attractive. Gravitationally speaking.

The supernova divorcee

Gravity is not the only cause of relationship breakdowns in our universe. Some relationships between stars end when one member of the pair has an outburst and sends the other away. Let's turn our attention to the case of US 708, a faint blue star who now wanders free from her partner after millions of years together.

She had a bright and glittering start to life as a hot massive star, creating chemical elements to make future stars and planets and adorning the night skies of thousands of planets with her piercing blue-white light.

She lived happily with her partner, a white dwarf. Although small, the white dwarf star was stable, caring and a great companion. But after a few million years, things started to change. US 708 got older, her energy levels grew lower, bags grew under her eyes and her waistline stretched beyond the orbit of her fifth planet. Her white dwarf lover changed, too. He became needy, stealing and feeding on the gas from US 708's outer layers to satiate his own energy requirements. The relationship had become co-dependent.

This situation was not sustainable for the white dwarf. His feeding became chronic, and the donated gas put podgy layers on him, making his heart strain. Eventually he collapsed and experienced a thermonuclear supernova. As he roared in his death throes, his partner was flung violently into space as the anchor to her world was removed and the invisible gravitational string snapped on her orbit.

Thrown into the Galaxy, she flew helplessly away from her former home, her partner and the only life she knew. Her new existence may not be exactly one she chose, but she has slowly come to love it as it is filled with adventure. Who wouldn't love backpacking around the Milky Way, visiting places rarely seen by wandering stars and meeting new stars and solar systems?

To this day, this fine blue star, US 708, soars through the Milky Way at 708 kilometres per second with her ears flapping in the breeze like a dog in a pick-up truck. Roaring past other stars and solar systems, she is admired and envied as the brilliant blue flash, a charming and charismatic star who appears like a great comet in the sky before moving on to the next quadrant of the spiral arm.

Does she recall her previous life with fondness and nostalgia, or is she too busy enjoying the ride? Maybe it's a bit of both.

The one that got away

The fastest star ever discovered in the universe is S5-HVS1. A fairly hot A-type star, he is soaring through space at 1755 kilometres per second. That's more than 6 million kilometres per hour!

His name betrays what is truly special about this star: his speed. The 'HVS' in S5-HVS1 stands for 'hyper-velocity star'. And this one is supersonic.

He was discovered in July 2019 during a routine survey of stars in a region of the southern sky. Scientists were looking in a part of the sky where there are lots of stars getting pulled by gravity into stretched-out stellar 'streams' between the Milky Way and other, small galaxies nearby. This star looked different from all the others, because his fingerprint of colours showed that he was moving extremely quickly.

Checking with the *Gaia* space telescope, which measures the precise positions of stars over consecutive months, we can see that the star is currently a breathtaking 29,000 light years from Earth and getting further away every day. If we trace his incredible journey back some 5 million years, S5-HVS1 almost certainly originates from the very centre of the Milky Way.

If you imagine the Milky Way as a flat pancake with spiral arms, this star is now whizzing up above the disc towards the spherical 'halo' of our Galaxy, with no sign of slowing down. Eventually, he will leave the Milky Way and live out his days in the isolation of deep space.

How did S5-HVS1 get onto this breathtaking trajectory? And what monster lurks in the middle of the Milky Way that flung our heroic interstellar traveller towards the edge of the Galaxy? Was it a supernova explosion that fired him out like a cosmic cannon? Was it a gigantic cluster of massive stars that caused the star to swing out of the region via a stellar partner transfer under the influence of a massive gravitational field?

Neither of these processes even comes close to explaining the unparalleled speed of the hypervelocity star S5-HVS1. What accelerated him to such an eye-watering velocity must have been something with superstellar power. Deep in the heart of the Milky Way lies a dark force so powerful, so mighty, that it can toss stars out of the Galaxy as if they are toys thrown from a playpen by an angry toddler. But what is it?

It's called Sgr A* (pronounced 'Sagittarius A-star'), and it's a supermassive black hole, 4 million times more massive than the Sun, that has been growing for billions of years as it greedily devours stars and interstellar gas.

The gravitational pull of a black hole is only the same as a star of equivalent mass, but have you ever met a 4 million solar mass star? You might be safe at a distance – many stars orbit serenely about Sgr A* – but stray too close and you'll be hit with an overwhelming gravitational force field. Stars that wander too near the edge are not just pulled into the black hole – they are pulled apart limb-from-limb as they fly screaming into the abyss.

Some stars have a narrow escape. They enter the danger zone as part of a group but, tragically, exit alone. Like a lion hunting a herd of gazelles, the black hole is ruthless in its efficiency. It picks off the weakest star and captures it, tearing apart its flesh and eventually consuming its gas. Its companion is left suddenly without an anchor, and the orbital momentum built up is released with a tremendous snap. The surviving star is hurled away from the scene, as if by a galactic catapult, at 8000 kilometres per second.

As it speeds away from the crime scene, it is clawed back by the black hole's enormous gravitational field, but the momentum of the star is too great to stop. Our survivor skids out through the Galaxy's crowded inner region, then out of the disc and into the Milky Way's more sparsely populated halo. As the star evolves into a red giant, its gas is stripped down to its bare bones. Finally, as a white dwarf it enters intergalactic space, a largely dark and lonely place, where it lives out the rest of its days.

An Englishman in New York

If you've ever travelled to a strange city or another continent, you'll know the feeling of entering an environment where the people, culture and language seem alien. I've experienced that most acutely in China, Korea and Japan, where I can't decipher the script or understand the language, and every street sign and menu is a mystery. In many nations, foreigners are described as 'legal aliens' despite not having tentacles or green goggly eyes.

But people travel to 'alien' places all the time, with about 40 million commercial airline flights connecting the world every year. The idea of stars travelling to alien places might seem extraordinary, though. Imagine a star taken from its home galaxy and wandering wide-eyed across deep space and into another galaxy. Is it even possible?

As we learn more about the interactions between galaxies, we are finding that intergalactic travel is not science fiction after all. Perhaps as many as half of all stars exist in transit between galaxies. This story of intrepid cosmic wandering is being told by a new generation of spacecraft, including the *Gaia* mission, which has been wildly successful at tracking the motions of stars through the Galaxy.

A recent study using *Gaia* looked at more than 7 million individual stars and followed how fast they were moving through space, and in which direction. In tracking their tiny motions, most were found to be cruising through the Galaxy's disc as they are supposed to, following the traditional orbital path taken by our Sun and the rest of the sheep.

There were, however, a few exceptions. Twenty of the stars studied seemed to be fast-moving free spirits travelling at several hundred kilometres per second relative to the Earth. These stars were treading their own path. How did they get up to speed?

Surprisingly, none of these speedy stars originates from the centre of our Milky Way, where the supermassive black hole Sgr A* resides. It seems that our resident black hole is not in the regular habit of pinging innocent victims into the abyss.

Seven of the twenty rapidly retreating stars were found to come from the disc of the Galaxy, likely flung out by orbital interactions in massive clusters. The remaining thirteen high-velocity stars found in the *Gaia* survey don't seem to have originated from our Galaxy at all. In fact, they're heading towards us from completely the wrong direction. These stars are extragalactic visitors – aliens, flying in for an exotic holiday in an alien environment, the Milky Way galaxy.

Given the recent nature of this discovery, astronomers haven't had much time to monitor the motions of these visitors and figure out exactly where they have come from. For example, one star has a velocity of 700 kilometres per second and lies more than 80,000 light years above the galactic disc. The most likely explanation is that it (and perhaps the other visitors) came from one of the dwarf galaxies that orbits the Milky Way, perhaps the Large or Small Magellanic cloud.

As these tiny galaxies orbit the Milky Way, they get stretched and shredded by gravitational interactions and many of the stars are extruded into space like kneaded pieces of dough. Studying the fingerprints of elements encoded in the light from each star may go some way to figuring out where they came from. That's because each dwarf galaxy has its own unique chemical mix, which is encoded in the make-up of the stars that are born there.

This extraordinary fact – that travelling stars exist – raises some thought-provoking questions. Where do the stars come from, and how many in our Galaxy are in fact immigrants from another land? This will be the focus of studies over the coming years, as will trying to piece together the likely path of the stars from their origins to the present day. As our telescopes get bigger and more powerful, we expect to discover more intergalactic wanderers from faraway lands.

11

Butterflies

Butterflies are famous for good reason. Their brilliant colours and the artistic symmetry of their delicate wings are almost unparalleled in the natural world.

It is well known that butterflies come into being through a tremendously complex process of metamorphosis. From humble beginnings as an egg, they transform into a caterpillar and slowly grow. The fully mature caterpillar retreats into a chrysalis and gestates there for several weeks before finally emerging as a delicate and colourful winged reincarnation of its mother.

Think about it for a minute. It's miraculous. What's even more astonishing is that exactly this type of evolution and rebirth is being repeated in trillions of stars across the universe.

The progression is similar in many respects. From a humble cloud of gas, a dwarf star emerges and slowly grows. It forms a cocoon, with its burning core gestating inside, before emerging as a beautiful nebula made up of glowing shells of colourful gas that resemble the perfect symmetry of a butterfly against the blackness of space.

The astronomical version of this stunning arthropod is called a planetary nebula. The name is somewhat misleading, since the clouds of gas left over from the expansion of a former red giant have nothing to do with a planet. It's just that astronomers in the 1800s thought they looked a bit like faded planets – big and round – and the name stuck.

A planetary nebula is a truly amazing stage in the evolution of a star. When we find one, we catch a fleeting glimpse of a rare creature that lasted only a few tens of thousands of years against the backdrop of the star's total lifespan of 10 billion years. If this were a human lifetime, the planetary nebula would exist for only forty-two minutes. That's how fleeting this transformational phase of a star's development is.

Despite their rarity, we see an array of these astronomical beauties in our skies. Their shapes are artistic and varied. Their colours are vibrant, although human eyes can't make out the vivid palette because our colour-sensitive optical receptors work very poorly where faint light is concerned, and the black and white sensors in our eyes go into overdrive.

This is remedied by using a telescope and digital camera with multiple colour filters, which are able to bring out the vivid hues hidden from our eyes. With a modern telescope, a planetary nebula is truly one of the most magnificent sights we can see.

A planetary nebula is a fleeting footprint of the irreversible transformation of a star into a new chapter of its life. Just where will this incredible transformation take the star next?

We can find out as we meet the butterflies.

Relight my fire

Ah, the Sun. Our life-giving star warms our planet, grows our food and drives our weather, but we don't like to think about what lies ahead for her. After all, our very survival depends on her being stable, unchanging and reassuringly boring.

I'm sorry to be the bearer of bad news, but she won't be that way forever.

Our Sun, like all of us, is a dynamic creature who never stops ageing – even though on the timescale of a human lifetime she barely seems to change. Although she is currently in a stable period of her life, some stages of her evolution will progress much more rapidly. The planetary nebula phase is one of those.

The transformation of our Sun from relatively boring to fabulous will begin in about 5 billion years' time. After steadily burning hydrogen for almost 10 billion years, this is when she is likely to run out of her regular fuel. It signals the end of her adult life.

At this point, she will swell up to form a red giant. This bloated behemoth is created when there is no longer any outward radiation pressure to counteract the inward force of gravity, which wins over quite suddenly and will cause our star's core to collapse.

As the inner part of the Sun crunches together, it will heat up, causing a shell of previously unburned gas around the stellar core to ignite, fusing hydrogen to helium; like the old Take That song, a case of 'Relight My Fire'.

The tremendous heat from this new layer of nuclear burning gas will heat the outer layers of our star and they will expand, causing her to balloon to hundreds of times her previous size. Although we can't be sure, she will most likely grow to engulf Mercury, Venus, Mars and possibly the Earth. (Bloody hell!) As she swells, her gas will cool down and change colour from a sweaty yellow to a cool, glowing orangey red.

What will happen next is mired in some mystery. The new hydrogen burning in the core of the Sun will begin to drive a gigantic stellar wind, an enormous sandstorm made from particles of dust that will form from the gases blown off her surface. This will create a 'snowplough' effect, driving all the gas from her nebulous outer regions.

What will transpire is the release of the cocoon that previously shrouded our red giant Sun in mysterious isolation. Her existing ravenous, burning core will become a white dwarf, no longer growling with nuclear reactions but merely glowing with radiant heat fuelled by the memories of her glory days. The glowing red embers of the outer Sun will flow rapidly into space and be heated by the searing radiation of the now-naked core of the inner Sun.

A planetary nebula, expanding into space. A tiny remnant core of the once-omnipotent star that drove the creation of the dinosaurs and our varied human civilisations.

The smoke ring

I started stargazing regularly in 1992, when I was twelve years old. I never had a telescope, but I used an old pair of binoculars to get a closer look at things like star clusters, planets, galaxies and clouds of gas (nebulae). I had a 1952 copy of *Norton's Star Atlas*, whose many pages chronicled the entire canvas of stars visible from the Northern Hemisphere, where I lived. It had a musty smell, but its pages were like a secret code that unlocked a whole universe of mysteries.

One of my favourite things to look at were the Messier objects – a bunch of faint, fuzzy things that a French astronomer, Charles Messier, chronicled about 250 years ago. He wasn't interested in the objects per se, but he was an avid comet hunter and wanted a list of fuzzy things in the sky that *weren't* comets but looked like they were, so he didn't get confused and rush off claiming a new discovery.

One of the most interesting of Messier's objects to look at through a pair of binoculars is M57 in the constellation of Lyra, the harp. It lies close to my favourite star, Vega, which always shone in my hometown sky like a magnificent white jewel. Next to Vega are two fainter stars called beta and gamma Lyrae. If I pointed my binoculars smack-bang between the pair, I could just make out the Ring nebula.

It didn't look like much through binoculars (a fuzzy blob, at best), but when I graduated to a telescope, borrowed from a nice man at the Braintree Astronomical Society, I could start to see its shape. The Ring nebula has a symmetrical circular smudge of white light hollowed out into a mysterious dark middle section, like the pupil of a human eye.

With a large telescope it is possible to make out a tiny point of light in the middle of the pupil. This is the white dwarf – the tiny, dense core of the former red giant. Previously enshrouded in her cocoon of gas, the core of the star didn't simply let her protective cloak float away aimlessly into space. Rather, she owned it. Like a sassy 1940s film star she formed her lips into a perfect 'o' and blew it like a smoke ring into the cosmos.

This is a statement nebula, created by a white dwarf star announcing her presence on the stage. As the unwanted gases from the previously bloated red giant floated into space, the newly emerged white dwarf was left naked and exposed, yet surrounded by an expanding cloud that revealed her great inner strength: the power to expel about 1.2 million trillion trillion tonnes of gas into the cosmos in an unbroken circle.

Such symmetry in space is perfection.

Imperfection is very common

Aside from the rare and haunting beauty of the Ring nebula, it is hard to find a perfectly spherical cloud of gas in space. The observed shapes of planetary nebulae are very puzzling: most of them (about 80 per cent) are in some way squashed or stretched, rather than spherically symmetrical like the Ring nebula. Many have complex shapes, with rings inside giant gas bubbles, or bipolar shapes like an hourglass.

Scientists, being curious about how everything works, started to wonder how the stars at the centre of these nebulae managed to blow bubbles with such complex shapes. After all, if a star is spherical (which they all are) then surely it would take a lot of effort for it to create a long or complex nebula.

One school of thought is that bipolar planetary nebulae are shaped by magnetic fields. This theory has guided much of my own research into the shapes of these complex objects.

As we know, the Sun is a magnet. With all the atoms broken up inside it, electrically charged particles swirl about and create a strong magnetic field that causes sunspots and guides violent solar flares into the darkness of space.

Galaxies are magnets, too. We are not entirely sure how the magnetic fields in these gigantic swirling cities of stars are generated – whether they are amplified from tiny 'seed' fields that are almost as old as the Big Bang or created from scratch by some complicated astrophysical process. What we do know is that they follow the galaxies' spiral arms and spill out into deep space. There is no escaping cosmic magnetism.

How these fields are able to shape and mould astronomical objects like planetary nebulae is something I've been interested in for a large part of my research career. What seems most likely is that although the star itself is spherical and symmetrical, the stellar wind that drives material from it is not symmetrical. The gas that is being blown off the equatorial regions of the star moves far more quickly than the gas that is leaving via the polar regions. This is one (very convincing) explanation for the weird and wonderful shapes of these beautiful cosmic butterflies.

The other possible explanation for their complex shapes is that the stars creating a planetary nebula are in couples or multiple-star families. Perhaps the gravitational pull of the second star can 'shape' the escaping gas, using gravity as its guide.

These intriguing interactions between stars, their ejected gas and the invisible forces of gravity and magnetism in space are still the subject of intense study. That's just as well, since some of the most inspirational images from our cosmos have come from studying this fleeting stage in the life of an ordinary star.

The menopausal star

When we study objects in outer space, one thing is almost guaranteed: as a rule, everything we look at is *really old* and *really, really slow to evolve*. Nothing much changes in the sky.

Or does it?

In 2016 I received an email from a retired Indian astronomer, Mudumba Parthasarathy (Partha to his friends). I had never met him, nor was I familiar with his research, but he had a proposition for me. He was an optical astronomer and had spent his career studying images of the sky taken with telescopes that gather regular light – the type we can see with our eyes. I am a radio astronomer, meaning that I look primarily at invisible stuff emitted in the form of radio waves.

His passion was the study of planetary nebulae, and in 1993 he had discovered the youngest planetary nebula ever seen by human eyes. It is called the Stingray nebula, and it probably began shedding its gaseous cloak in the 1980s. Since then, astronomers have studied the shape and structure of the gas and watched as the temperature and chemical properties of the star have changed, using images from various large telescopes.

What Partha wanted from me was my radio astronomy expertise. The weird behaviour in the central stars of the Stingray nebula had been observed several times since 1991 using an Australian radio telescope, and there were a large number of images languishing in that telescope's electronic archives. Could I find the time to process these images and analyse the results, he asked, to shed further light on how the star had evolved over the past thirty years?

I jumped at the challenge, and worked with a team of colleagues to process the images. And I requested further observing time on the Australian radio telescope compact array near Narrabri in New South Wales to study what the star was up to now.

What we found was incredible. Since 1971, when the star was first measured, it has experienced a menopausal hot flush called a 'late thermal pulse', where a flash of nuclear fusion took place in its centre, causing it to heat up. This event drove off large amounts of the gas from the outside of the star into space, making it glow.

A planetary nebula was born. With my latest radio images, along with some optical pictures from the Hubble Space Telescope, it was clear that the gas surrounding the central star had complex shapes, including a bright ring inside a larger bubble, with a pair of 'ears' to complete the picture. The Stingray nebula looked like a character from the board game Guess Who.

Significantly, the colours of radio waves from the ears may herald the start of a large outflow of gas from the magnetic field of the central star. The larger ring structures in the Stingray nebula were possibly caused by the gravity of the star that orbits the central star (it's a binary) aligning the gas into such a shape, but we can't be sure.

All of these factors (the temperature of the star and the size and shape of the nebula) have changed each time we have studied the Stingray nebula. That we have detected such enormous changes to the star and its gaseous envelope within a human lifetime is simply thrilling.

The star is currently cooling after its outburst, but will such an outburst happen again? And how long will its menopausal symptoms last? What is next for the Stingray nebula?

The only way to find out is to keep looking up.

Put a ring on it

As an astronomer, I am drawn to the natural beauty of the cosmos. The attractive swarm of cosmic butterflies that flits through the spiral arms of the Milky Way has to be the most enchanting cohort of astronomical insects in the entire universe.

It's no surprise, then, that I was inspired to study these complex objects, to understand how they grow, what influences their shapes and what role they play in the evolutionary lives of stars.

One of the most exquisite planetary nebulae in the Milky Way has to be the Saturn nebula. It is named after the giant ringed planet in the solar system that we all know and love. With its spherical halo, bright shells of gas, huge 'ears' and long, narrow jet of material streaming from the central star, the Saturn nebula gives us a glimpse of what the Stingray nebula might look like in a thousand years or so.

The central star of the Saturn nebula is a white dwarf with a surface temperature of around 55,000 degrees Celsius. The strong ultraviolet radiation from this star causes oxygen atoms within the nebula to glow with a bright-green light that is obvious even to backyard observers through a small telescope. With modern telescopes, high-resolution colour images show a range of exciting structures and clouds of dust. All of this information is being painstakingly pieced together by astronomers trying to recreate the history of this unusual star.

Some of the weird shapes in the nebula might be caused by the uneven shedding of gas as the giant star that created the nebula huffed and puffed in giant stellar pulsations. Others might be driven by magnetic fields, or by a hidden binary star that we cannot see.

Even if its past is unclear, the future of the Saturn planetary nebula is certain. Like all of its kind, it is only visible because radiation from the central star is causing the surrounding gas to shine. Once the gas expands and dissipates, the nebula will fade from view forever. Just as terrestrial butterflies are marvellous at pollinating flowers on Earth, our cosmic butterfly the Saturn nebula has its job as a pollinator of our Galaxy. As the outer regions of the star are shed, chemicals including oxygen, carbon, hydrogen and even some metals will be flung into the cosmos, spreading the seeds of new stars and planets. In 10 billion years' time, it is possible that another life form made from these chemical elements, in a new solar system, will be reading a book about their humble cosmic beginnings.

That's the circle of life, which will hopefully continue for trillions of years to come.

12

Dizzy stars

Neutron stars sound like the stuff of science fiction. Made of a substance denser than almost anything in the universe, they can spin thousands of times every second. These enigmatic stars beam energy across the Galaxy at light speed, and their size matches that of a large suburb.

How are they made? Giant stars burn their fuel very quickly and explode as supernovae, splashing their gas across the Galaxy while only their tiny dried and shrivelled-up core remains. Under the influence of gravity, the remaining parts shrink into a tiny volume with a radius of 10 kilometres or so. That's right – an entire star is crammed into a space the size of a small town. As the core gets denser, the subatomic particles of gas get squished together and form a gigantic mass of neutrally charged matter, called neutrons.

That is how we make a neutron star.

The contraction of a neutron star can't go on forever. It quickly settles down when the repulsive force between neutrons, called the strong nuclear force, starts to counterbalance the attractive force of gravity. At that point, the neutron star relaxes into a very long stint in front of the telly.

Our tiny friend can't get *too* comfy, though, because something else happens when a gigantic star shrinks: it starts to spin very fast.

Have you ever seen an ice skater in the winter Olympics? There's a particularly impressive move where they spin around really fast, using their arms to build up speed. They bring their arms in towards their body and start to rotate even faster, until everything's a blur, no doubt for them as well as for us watching at home.

This works because of a physical law called the conservation of angular momentum. That's a fancy way of saying that if something is spinning and you make it bigger, its spinning slows down. If you make it smaller, it spins faster.

Neutron stars are formed when a star gets a lot smaller. Contracting a star from a few million kilometres across to about 10 kilometres across increases its spin rate enormously, from about once per month to around ten times every second. No wonder neutron stars are dizzy!

Will the Sun ever become a neutron star?

Not a chance. Our Sun is not massive enough to crush electrons and protons together and form a material where one teaspoon's worth weighs the same as Mount Everest. This particular party trick is left exclusively to the most massive groups of stars.

Neutron stars may not shine brightly anymore – their nuclear reactors have long since shut down – but they can be seen, simply because of their residual heat. Their sheer diversity and unconventional behaviour make them eligible for nomination as the most peculiar stars in the universe.

So what sort of housemate would a neutron star be? Let's go on a dizzying journey into the hidden world of neutron stars.

Pulsars

Pulsars are those annoying flashy types.

Unlike regular neutron stars, who largely keep themselves to themselves, pulsars seem to like to signal their presence right across the Galaxy.

A pulsar is simply a neutron star that flashes on and off very rapidly. That's why it is called a pulsating star, or 'pulsar' for short. It shines brightly in radio waves, not light, which gives us clues to how it works.

Astronomers figured out fairly quickly that pulsars aren't really twinkling like the Christmas lights I still have hanging outside my house even when it's nearly May. Rather, they are rotating very rapidly and beaming two spinning searchlights towards Earth, like a lighthouse. To us, this beaming column of light looks like a flash.

Pulsars are created in the immediate aftermath of a supernova explosion, and some, like the Vela pulsar, are still surrounded by the dregs of gas left over. We call the expanding gas from this event a supernova remnant.

The Vela pulsar was created in a supernova explosion about 12,000 years ago and is still spinning almost as fast as the day it was born. Images from NASA's Chandra X-ray telescope show how the intense radiation from the pulsar is illuminating a nebula surrounding the star, leaving a trail of hot plasma streaming away from its poles.

Vela also has a party trick. Every few years its rotation slows down ever so slightly, then speeds up again. It's the pulsar equivalent of a hiccup. Astronomers think this might have something to do with the breakdown of vortices within different layers inside the star, but we can't be sure because it's impossible to get close and take a look.

Venturing anywhere near a pulsar would be intense, to say the least. Temperatures would be in the hundreds of thousands of degrees. Scoop up a teaspoon of pulsar material and it would weigh a billion tonnes. And gravity would be 2 billion times stronger than it is on Earth.

All in all, pulsars are probably the strangest stars we know.

The double pulsar

When I was studying for my PhD at Jodrell Bank Observatory in the UK, there was a friendly rivalry between those astronomers who were studying pulsars and the rest of us.

The pulsar astronomers would have their own meetings and speak their own language, with jargon like 'p-dot' and 'spin-down rate' flying around. I can't remember half of what they were saying because, frankly, it sounded like gobbledygook. They were adamant that pulsars were the best thing since sliced bread, but we used to tease them, saying they didn't understand how pulsars worked and they were just 'stamp collecting'. Ah, those were the days.

Pulsar astronomy began when Jocelyn Bell Burnell discovered the rapidly spinning stars when she was a graduate student at Cambridge University. This outstanding achievement was awarded the Nobel Prize – but not for her. Bell's supervisor, Antony Hewish, and another man, Martin Ryle, received the honour and a US$124,000 cheque for developing the technology behind the discovery in 1974. That was a lot of money at the time (equivalent to more than US$1 million today).

More than 1000 pulsars have been discovered in the intervening years, including pulsars in long-term relationships with other stars. Astronomers got very excited at the prospect of using them to measure the fundamental rules of the universe, including the passage of time and the laws of gravity.

This idea got even more exciting in 2003 when Marta Burgay, an Italian radio astronomer, discovered the first (and still the only known) double pulsar. Pulsar A spins every 22 milliseconds; pulsar B revolves rather more slowly, once every 2.3 seconds.

A pair like this experience very strong and tangled gravitational fields. Since the spinning stars act like exceedingly accurate clocks, we can use deviations in the regularity of their flashes to test our understanding of how gravity works according to Einstein's theory of gravity.

The stars in the pulsar pair are very close to one another, orbiting once every 2.4 hours. As each drags through the enormous gravitational field of the other star, it loses energy, a bit like walking through deep sand. Gravitational waves are released, which reduces the orbital energy of the pulsars and pulls the stars closer by about 7 millimetres per day. We can't see the gravitational waves, but since the stars' flashing is changing we can measure very precisely that the predictions of Einstein about these disturbances in the fabric of space and time were extremely accurate!

Subtle changes to the timing and intensity of the pulsar flashes showed astronomers that the orbital angles of the pulsars were changing. The lighthouse beams of star B have already twisted out of view, but they are predicted to return to our line of sight in the year 2035. Due to the loss of gravitational energy, the two stars will collide and coalesce in about 85 million years' time.

These relics of stellar giants are tiny and extreme and have much to teach us about the underlying forces in our universe. I guess pulsars are interesting after all.

Magnetar

Imagine a magnet so strong that it could wipe a billion credit cards at a distance of 180,000 kilometres. A magnet 1000 trillion times stronger than the puny one generated inside planet Earth. A magnet so powerful that it would stretch atoms into long, thin tubes and rip your body's molecular bonds apart, transforming you into an atomic version of raspberry jam in the blink of an eye.

A weird and scary breed of neutron star displays these very properties. They are called magnetars, since they are the most magnetic objects in the universe.

Magnetars are a type of neutron star that exhibits very fast rotation and the eye-wateringly strong magnetic fields with all the properties described above. The last few astronauts we sent to study them were turned into a messy human paste – I'm joking! We don't know exactly how they generate their enormous magnetic powers. Most astronomers studying them (from a safe distance) think they probably come about when electrically charged particles like protons swirl around really fast in the star and create a dynamo effect, just like the swirling of charged particles inside our own Sun.

Conditions inside a magnetar are about as far from those of the Sun as you can imagine, though. The surface temperature of a magnetar is a few million degrees, compared to just under 6000 degrees for the Sun. Magnetars spin as fast as the blades of your kitchen blender, whereas the Sun takes twenty-four days to amble around once on its axis, surveying its surroundings with all the urgency of a retired holidaymaker. The Sun is big and fluffy, whereas a magnetar is denser than any known substance in the universe. Not to mention its magnetic field, amplified exponentially as it is crushed into a tiny space no larger than Oxford's city centre (that's small, by the way).

You might wonder how a star made of neutrons, which do not have an electrical charge, generates a magnetic field. The answer is that it doesn't. There are still plenty of protons and electrons lying about in a neutron star, especially towards the surface where the crushing pressure is not so great. The interior of a magnetar is probably in a superfluid state. That means the spinning material inside the neutron star has essentially no friction and will keep spinning forever! That is in stark contrast to the surface of a magnetar, which probably has a solid crust that is locked onto the magnetic field.

When magnetars experience starquakes, which break or fracture the crust of the star, they can create tremendous bursts of the most energetic radiation possible – gamma rays and X-rays – that are visible across the universe, even in other galaxies. Around 50,000 years ago, a magnetar called SGR 1806-20 experienced a starquake that released a burst of energy reading 10,000 trillion trillion trillion watts and lasted for only one-tenth of a second. In that time, the magnetar released more energy than our Sun has emitted in the last 150,000 years.

In December 2004 this wave of radiation finally reached the Earth and began to register on the gamma ray and X-ray detectors at dozens of telescopes around the world. It even managed to overwhelm some instruments, with radiation measuring off the charts. Perhaps most incredible is that the wave of high-energy radiation from this distant neutron star zapped our Earth's atmosphere and temporarily increased the size of the ionosphere (the Earth's ionised atmospheric layer) by 25 kilometres.

This event was one of the biggest explosions ever seen in the universe. It broke up atoms in a large part of the Earth's upper atmosphere. And it all came from one of the smallest astronomical objects we know – smaller than the Earth, the Moon or even Halley's comet.

Beware the powerful nuclear forces lurking inside every atom, just waiting to be unleashed.

Black widows

You may have heard of black widow spiders?

The Australian redback is a particularly venomous example of these dangerous arachnids, which possess a powerful chemical that can dissolve an insect in seconds. The spider injects it with a mix of digestive enzymes and sucks up its middle (yum).

Although they are not overly aggressive, black widows sometimes bite humans when we accidentally sit on them, put our foot into a shoe in which they have made a home, or otherwise bother them in their natural environment. Without medical treatment, their venom is capable of killing an adult human within hours.

The name black widow reflects the mating habits of the female spider, who sometimes eats her suitor immediately after mating, or sometimes even during or before. Don't ask me why they do this, but perhaps she knows her own power. The female of the species is substantially larger than the male and clearly knows what she wants.

Like many inhabitants of planet Earth, black widow spiders have an equally scary astronomical analogue.

In 1988, a rapidly rotating pulsar was discovered with another star in close orbit. The pulsar spins more than 600 times per second (just imagine that for a moment) and its companion is fairly small, a brown dwarf star. The pair orbit one another every eight hours, with the brown dwarf regularly eclipsing our view of the pulsar each time they cross paths.

If the pulsar is the female spider, the dull and squat brown dwarf is the male. As they dance together in close contact, she controls his every movement. Their eyes are locked onto one another as the pulsar controls their rapid spin from a distance equivalent to only 1 per cent of the distance between the Earth and the Sun. Gas is exchanged and the pulsar weaves the brown dwarf into a cocoon of his own gas. Her gravitational dominance makes sure that the male partner is slowly cannibalised by her.

Between the stars, the exchange of hot material drives bow shocks – collisions of hot, ionised plasma – creating bursts of high-energy radiation in gamma rays and X-rays. This is how this dance has been seen across the cosmos by astronomers, who are equally terrified and appalled by the abominable spectacle. It surely is a star-eat-star cosmos out there.

Quark stars

If you studied any physics at school or beyond, you may be familiar with the constituents of atoms: protons, neutrons and electrons.

These in turn are made up of quarks, of which there are six flavours: up, down, top, bottom, strange and charm (I'm not even kidding). Protons and neutrons are each made up of three quarks. Electrons are different and are not made of quarks.

There are many other subatomic particles, including the Higgs boson, neutrinos and other force-carrying tinies, which can be released in high-energy collisions in research facilities like the Large Hadron Collider, where particles are smashed together at extraordinarily high speeds. We don't normally encounter these in the 'real world' of our everyday existence, but they are tremendously important in building everything we see.

This magical quantum world has been uncovered by the study of particles in extreme environments both on Earth and in space. The mind-blowing thing about particle physics is that often, the existence of subatomic particles is predicted using mathematics long before they are discovered in the flesh. Eventually, though, these theoretical particles show up in colliders or in astronomical observations from exploding stars, or in hot, colliding interstellar gas that betrays the underlying nature of the matter we take for granted every day.

There is, however, a type of star that remains 100 per cent theoretical. We have never seen such a specimen in nature. The quark star is a hypothetical star whose density exceeds that of white dwarfs and even neutron stars. It's a star that exceeds the maximum weight of a stable neutron star but manages to contract further and stabilise as a bundle of quarks.

Although quark stars are a theoretical proposition only, the idea follows a pattern of many astronomical discoveries predicted by physics theory and then later discovered in the real world. Nobody knows how this entirely new and hypothetical type of matter might behave. It could follow weird and magical ways that we can't even begin to imagine.

Astronomers are curious creatures always on the lookout for new evidence of weird or unusual objects in the universe. These strange beasts, if real, would add an ethereal new dimension to the cosmos as we know it.

Black holes

Throughout my career I have visited hundreds of schools to talk about my job as an astrophysicist. One of the most exciting parts of talking to kids is when we get to the Q&A session. The questions always flow in the same way.

'Have you ever been into space?'

'Er, no, but I'd like to!'

'How many stars can we see in the sky?'

'Around three thousand if it's really dark.'

'What's a black hole?'

'Well …'

This final question is the most common one that kids ask. According to just about everyone, black holes are the most mysterious, exciting and thrilling astronomical objects in the universe.

They are also the most misunderstood, with many people thinking they are dangerous and violent (which is partly true) and suck up anything within a huge radius, like a giant vacuum cleaner (which is not really true).

The truth is that black holes are no different from ordinary stars when it comes to gravity. If tomorrow you replaced the Sun with a black hole that weighed exactly the same, the Earth and all the other planets would continue to orbit unchanged. It would just be very dark.

Black holes are a muddle of contradictions. They are descended from stars but don't shine like stars. They are infinitely small and dense, yet their sphere of influence can be as big as our solar system. They emit no light, yet they are among the brightest sources of radiation in the universe because of the sheer volume of hot gas from ripped-up stars being dragged across their threshold kicking and screaming.

A black hole, to all intents and purposes, is a completely dark star surrounded by a no-go zone, with a central core that has completely disappeared up its own backside. It's as simple as that.

For those of you who are still confused, let me shed some more, ahem, light on these mysterious dark creatures.

Stellar-mass black holes

At the end of a massive star's life, it runs out of fuel to burn. The V8 engine that has roared with the ferocity of a nuclear bomb for more than 10 million years slumps into calmness, silent and dead.

Gravity has been pulling this star together for eons, keeping it from exploding. Now it is gravity that wins, pulling the core of the former star into ever-decreasing dimensions.

What happens next depends on the initial size of the star. If it started out like our Sun, the end of nuclear burning will see its core sink down into a white dwarf, settling at a little over the size of the Earth.

More massive stars (1.4 to 3 times the mass of the Sun) are too heavy to be supported by 'electron degeneracy pressure' (the force of electrons not wanting to pile on top of each other). The gravitational pull is too strong and these spent stellar cores sink even lower, around 20 kilometres in diameter, into a neutron star.

For stars a little heavier still, a gravitational pile-on has catastrophic consequences. With so much weight, the core of the star buckles under the pressure. Electrons can't stop the collapse, neutrons can't stop it, and the core melds and mushes together into an infinitely small point: a black hole singularity.

This tiny stellar relic, smaller than a pinhead, is surrounded by warped yet almost empty space bordered by an invisible edge called the 'event horizon'. This is the point of no return. From inside this boundary it is impossible to escape. Outside this imaginary line, space and time are curved and warped in such a way that anyone hovering near the edge would experience time travelling far more slowly than they would on Earth.

I know. The whole thing sounds ridiculous. Surely it can't be real?

I would bet my house on it being real. Since the early 20th century, every prediction of the theory behind black holes has been observed in reality.

We have found black holes cannibalising their partners, who scream with blood-curdling tones as they spew X-rays into the cosmos. These are called X-ray binaries.

We have seen black holes at the centres of galaxies, their deathly stellar waiting rooms filled with hot gas that hurls radio waves across billions of light years from the furthest reaches of the cosmos. We call these active galactic nuclei.

And we have felt the vibrations, the heavy footfalls, of distant black holes combining together in galaxies far beyond the Milky Way. These are called gravitational waves.

Not to mention the recent triumph of the very first image of a black hole.

With overwhelming evidence like this, we have no doubt these mysterious monsters are real. The only question is, will we ever understand what's going on inside their impenetrable borders?

Our friendly neighbourhood supermassive black hole

The remnants of collapsed massive stars are not the only flavour of black hole we see in our diverse universe. Just like stars, these cloaked enigmas come in a vast range of different sizes.

Black holes that come from the cores of dying stars have masses of a few times that of our Sun and radii of a few kilometres (that's the size of the event horizon – the boundary from which light can't escape). Deeper in space, there are black holes that are millions of times heavier and larger than these dying stars. One such colossus can be found loitering in the middle of the Milky Way.

The first clue to the enigmatic beast hiding in our Galaxy's centre was a bright source of radio waves discovered in the constellation of Sagittarius in 1931. Radio telescopes had only recently been invented, and radio astronomers found that the sky was filled with interesting and unexplained objects that in many cases had no analogue emission in visible light.

The scientists realised that not only was the radio emission from Sagittarius very bright, it was also dense, compact and point-like, like a star. This was strange because most bright sources of radio emission were galaxies, which are very large and extended in size. Whatever this little firecracker was, it certainly packed a punch for its size.

It was also noticed by astronomers that the diminutive power-house of radio waves appeared to lie smack-bang at the geographical centre of the Milky Way. Coincidence? Maybe not.

The nature of the lively little object in our Galaxy's midst was the subject of much debate. It was named Sagittarius A* (pronounced Sagittarius a-star), the asterisk indicating that this radio source was of special interest. You might recall that we met Sgr A* in the 'Runaways' chapter.

Many years later, when the technology became available, telescopic infra-red (heat) cameras were turned to the centre of our Galaxy. Not much was found from the heat map of Sgr A* itself, but infra-red images showed a swarm of stars orbiting it. They were not visible in normal light, since this region is cloaked in dense and dark clouds of molecular gas that blocks pretty much all light.

A small cluster of six stars is locked to Sgr A* in a variety of elliptical circuits. Some of the six pass very close to it, as close as the orbit of Uranus around our Sun, and others venture further out before returning, like the orbit of a comet around the Sun.

This starry family was the silver bullet that confirmed the nature of this intense anchoring force at the centre of the Milky Way, because we know how stars behave when they are orbiting an object, even if that central mass is unseen. If we can calculate the mass of the stars, we can use their orbital sizes to calculate the mass of the large central object.

The answer was shocking. The central mass around which the stars were orbiting weighed 4 million times the mass of the Sun. The colossal heft was crammed into a region only 12 million kilometres across, and surrounded by a vortex of hot gas emitting radio waves, and also a family of stars.

It could only be one thing: a supermassive black hole.

Further out, clouds of gas acted as mirrors, reflecting 'light echoes' from previous outbursts from the region. Subsequently, a gigantic bubble was detected around this object at the centre of our Galaxy, also telling of a hot and energetic history.

The mega black hole may be fairly quiet right now, but in the past it has experienced significant events where it spewed high-energy particles, X-rays and gamma rays into the Milky Way. Astronomers calculated that the energy required to blow these bubbles (called Fermi bubbles after the gamma-ray telescope that discovered them) was equivalent to swallowing about 10,000 Suns' worth of gas.

To have a strapping bouncer guarding the centre of our Galaxy might be reassuring to some, but given the unpredictable nature of this gentle giant, it might be prudent to keep watching Sgr A* in case we get in the firing line of one of its outbursts.

Say 'cheese'! Oh, you blinked

If you flew in a spacecraft past a supermassive black hole, what would you see?

We know that black holes are bright because they are surrounded by gas from stars that have been ripped to shreds. No light escapes from the black hole itself, but by its violent virtues it attracts a lot of noise.

For a long time, we could only predict what a black hole might look like. What shape and impression its shadow would make against the incandescent background of searing-hot gas. How the stellar entrails being hoovered up by the black hole would make their last few laps before succumbing to their inevitable slip-and-slide ride into oblivion. Predictions were made using computer simulations of black holes as they consumed stars, like the portly King Henry VIII of England mopping up a chicken carcass.

It turns out that these simulations were uncannily accurate.

In 2019, a humongous global team of scientists released the first-ever image of a black hole. It was the supermassive black hole at the centre of M87, a nearby elliptical galaxy that is more than 16 million light years distant from Earth. The image was made using a network of radio telescopes in different parts of the globe, many perched on remote mountaintops. Collectively, they were called the Event Horizon Telescope.

Using many telescopes as a team is not a new technique. Radio astronomers have been employing this method since the 1940s; indeed, I have spent my career in this field. My first job was as a support scientist at the Joint Institute for Very Long

Baseline Interferometry in Europe (JIVE for short), which was the European hub for a network of more than twenty giant radio telescopes around the world.

To make an image of something very small and far away, we use a team of telescopes spread out over as large an area as possible. This increases the amount of detail we can make out – it's essentially creating a telescope the size of the Earth!

That's what the Event Horizon Telescope team did. They also tuned their telescopes to very high frequency radio waves, which further ramped up their detail-resolving power. After years of trying and a complex combination of data, the first image was released. It looks rather like a blurry orange doughnut. Check it out online if you haven't seen it.

Now let's unpack what we're seeing.

The hole is where the black hole sits as it mops up all the light that comes anywhere near it (within the event horizon, or point of no return). Ultimately, the bright orange represents radiation from ripped-up stars circling the black hole before the gas falls in. In this particular image we are looking at radio waves, not light. However, if you were nearby you'd probably see light too, so let's use that word to mean all radiation.

The inner edge of the doughnut ring is the 'light orbit' of the black hole. This light is not just coming from the sides of the black hole, but also from behind it. That's because the gravitational field of a black hole is so strong that it bends light around it. The orange circle of light has been bent right around the black hole, like a lens. Weird or what?

There is a bright 'hotspot' on one side, representing the spinning of the gas. Where the material is moving towards us, the light appears boosted. The brighter edge of the doughnut ring is moving towards us and the fainter edge is moving away from us.

This image of a supermassive black hole in a galaxy 54 million light years away is remarkable not just because of the sheer technical difficulty of creating it, but also because it is the first time human eyes have seen what a black hole looks like. That it matches our predictions gives us hope that our theories of the universe – and of the fundamental force of gravity – are on track.

Next comes the challenge of photographing Sagittarius A*. Although Sgr A* is closer to Earth, it is much more challenging to get a good shot because it rotates so much faster. As anyone with kids or pets will know, getting a good photo of a moving object is hard!

The missing link?

If you want to start an argument at an astronomy conference, talking about 'intermediate-mass black holes' is a great way to do it.

All black holes are greedy cannibals who ambush, dismember and consume stars that get too close. As they do so, they gain weight. As the gas falls over the event horizon, it enters a steep slope – the astronomical version of a skiing 'black run' that condemns the gas, now shredded into a stream of subatomic particles, to an eternity in the black hole's singularity.

Although stellar-mass black holes start their lives and grow very gradually as they eat, the origins of supermassive black holes are not well understood at all. We see them consuming gas – sometimes large amounts of it when galaxies merge. This is what powers so-called active galactic nuclei, stunningly bright galaxies that shine in radio waves and are ubiquitous throughout our universe. But how did supermassive black holes come about in the first place? Were they born many billions of years ago, in the beginning, when galaxies formed? Or did they grow gradually from smaller black holes over time?

One thing's for sure. Supermassive black holes did not grow from stellar-mass black holes. The supermassive variety are millions or sometimes billions of times more massive than a black hole that comes from a star. Small black holes simply don't eat fast enough to grow into the supermassive variety.

To investigate how supermassive black holes originated, we can look into what the middles of galaxies were like more than 10 billion years ago.

Did I just say that we can look back in time? Yes, I did!

It's possible because of a clever trick of the light – the speed of light, to be precise. If we look at radio waves coming from a very distant galaxy, let's say 10 billion light years away, the radio waves have taken 10 billion years to travel to us. That means we are seeing the galaxy as it was 10 billion years ago, when the light was released. A time machine, of sorts.

We can tell how massive a black hole at the centre of a galaxy is just by looking at the speed of gases moving in the centre of the galaxy. That enables us to figure out whether very massive black holes were around in the early days of the universe.

So what do we find?

Turns out that there were *heaps* of supermassive black holes very early in the universe's formation. Less than 700 million years after the Big Bang, there were galaxies with supermassive black holes that had already grown to hundreds of millions of times the mass of our Sun.

Where they came from is unknown. We simply don't have telescopes powerful enough to probe even earlier periods in the universe's development. Did they grow organically, albeit very rapidly, from black holes created from the massive cores of mega-stars in the early universe? Or did galaxies form with their own fairly massive black holes (perhaps a few thousand solar masses) already in tow?

Given a few million years, perhaps these giants had enough time to combine and grow into the gargantuan beasts we see today. Searching for the missing links, called intermediate-mass black holes, is a field of great interest to astronomers, who are still struggling to explain the origins of the galactic giants.

Neither model has yet won out, with circumstantial evidence for black holes a few thousand times the mass of the Sun, and clear evidence for those with 100,000 to a million times the mass of the Sun, hidden inside puny low-mass galaxies.

None of this evidence yet provides a clear timeline for the growth of supermassive black holes from the early universe to today. As telescopes and gravitational wave detectors improve, we will have many more tools at our disposal to piece together the formation of these important building blocks of our universe.

Sharing a sofa with tiny black holes

We've met big black holes the size of a solar system, and those the size of a small country town. You've heard about the middle-sized black holes, which could possibly be a stepping stone between the two. But what about teeny-tiny black holes – the type that could be floating around in the room with you right now. Could they exist?

Our universe began in a single point called a singularity. At the moment of the Big Bang, all of space and time was curled up into an infinitely small point. All the energy, the matter, the heat and light and joy and love of the universe was in one place, just waiting to be unleashed.

We have no idea how or why, but something triggered that singularity to expand. And it's just as well, because the ensuing chain of events led to the stunningly beautiful universe that we inhabit today.

Early on, conditions were hot and the environment was trying. Turns out that cramming the entire universe into a space far smaller than a pinhead makes it hot and angry. Atoms didn't exist, electrons and protons had not yet formed. All that existed was a sea of furious quarks, antimatter and other subatomic midges.

Within this dense and chaotic environment, less than a millionth of a second after the Big Bang, some cosmologists believe that the conditions could have created tiny versions of these dark hollows called 'primordial black holes'. The smallest of these, weighing less than 100 billion kilograms (that's about the mass of Mount Everest), may have floated everywhere throughout the universe for the first few billion years.

Even if they did exist, these tiny black holes will have evaporated by now, via something called Hawking radiation, a complicated process that allows energy to 'leak' from black holes even though no physical particles are escaping the event horizon. It's too mathematical to explain here (and many basic metaphors are just plain wrong when it comes to quantum mechanics), but if you want to do a degree in cosmology I think you'll find it fascinating.

A slightly larger primordial black hole – one that weighs a trillion kilograms – might be sharing the sofa with you right now. Black holes with masses up to about 1000 times the mass of our Sun may be observable. We see them because when they whirl around in tight orbits with neutron stars or other black holes, they release a spray of gravitational waves. Gravitational wave detection of black holes is the most exciting prospect for their study from Earth.

It is now of utmost interest to astronomers to count the number of black holes in the universe and figure out how heavy they are. This will help us to understand the processes by which they are generated, both in the post–Big Bang universe and by later stellar collapse. It will also help us uncover whether so-called 'dark matter' – the missing mass in the universe that doesn't emit or absorb light – could be composed of many small black holes, or whether there is another explanation.

Far from being science fiction, these strange and alien objects are now in the scientific mainstream, and learning more about them might just solve some of the biggest outstanding questions we have about our universe.

Goodnight

When I began my journey into astronomy as a teenager, I remember being continually amazed at how much scientists knew about our cosmos.

Prior to my first forays into the subject I don't think I'd really thought much about the lives of stars at all. The books I read taught me lofty facts about the glittering world that arches above our heads at night. And not being satisfied with simply believing what people wrote, I made a point of observing them too.

From my first books on amateur astronomy I learned that stars like Betelgeuse were gigantic and orange and that others, like the north polar star Polaris, were compact and bluey-white. To check this out for myself I learned how to de-focus my SLR camera and leave the shutter open for an hour so that I could photograph the fat trails of light these stars left as the Earth rotated on its axis. The star trail photographs I took enabled me to see their colours in vivid clarity.

More in-depth reading taught me that the majority of stars were not loners, as a casual observer of the sky might believe, but that they lived primarily in pairs or larger families. My explorations with a borrowed telescope from my friends at Braintree Astronomical Society confirmed this. From the frozen wasteland of my back garden, I witnessed firsthand that stars like Mizar and Alcor in the constellation of the Great Bear were a pair, each closely bonded to its partner. I watched as clusters of stars emerged from their pre-stellar cocoons of natal gas in stellar nurseries like the Orion nebula.

I discovered how a supernova explodes when a massive star nears the end of its life. I have never seen a supernova in the sky, but in my professional life I have studied the remnants of long-dead stars that exploded thousands of years ago and whose remnants are still expanding into space. Using radio telescopes and X-ray satellite images, I pieced together the history of the explosions like a forensic scientist, unravelling how the strange shapes of the resultant nebulae were influenced by the magnetic field of the Milky Way.

My university studies in astronomy and astrophysics taught me that a star nearing the end of its life can experience a sudden flash of helium burning and be born again. In doing so, it throws off its outer cloak of gas and bathes its surroundings in a warm ethereal light, becoming a planetary nebula. It's somewhat like the adventures of a grey nomad in a flurry of exploration and travel after many years of the constraining conformity of working life, planning new adventures to round off a long and happy life well lived.

A decade later, I used the six giant radio antennas in Narrabri, New South Wales to study the transformation of the evolved blue star SAO 244567, which was undergoing its own 'hallelujah' moment. I watched with nervous excitement as the star heated by 40,000 degrees Celsius and threw off its outer gas to form the Stingray nebula. As I created the first radio pictures of the nebula, I bore witness to a many-billions-of-years-old star settling into its long and quiet future as a white dwarf.

A clear and inky black sky is one of the greatest joys I have ever experienced, and I hope I will spend many more years marvelling at its pure beauty as well as uncovering its scientific wonders. The secrets its infinite caverns hide can inspire us all.

Stars are a weird and unpredictable bunch, so you never know what you might witness or observe next. Don't take my word for it, though. Go and see for yourself.

Acknowledgements

To Sally Heath, Sam Palfreyman, Katie Purvis and the whole team at Thames and Hudson Australia, thank you for lending your expertise and passion to make this book a vivid reality. What motivates me to write about the universe is not anchored in the stars at all; it is predicated on the people here on this fragile Earth and their passion for looking beyond themselves towards something greater.

To my family, as always, I am in debt to your love.

Further inspiration

Websites

https://www.space.com – A great source of international astronomy and space news, including articles on how to get started with stargazing.

https://www.NASA.gov – The United States of America's National Aerospace and Space Administration website. Contains amazing images, videos, news stories and live streams covering the US human spaceflight program, astronomy and robotic exploration of the solar system.

https://www.esa.int – The European Space Agency website, containing information about Europe's space missions, science and technology.

https://spaceaustralia.com – Australian space news, articles and podcasts. Great for southern hemisphere observers.

https://astronomynow.com – United Kingdom astronomy news, articles, information and tips for amateur astronomers.

Smartphone apps

Just download one these apps onto your smartphone and hold it to the sky to automatically find constellations, planets, satellites, meteor showers and more!

Sky Guide – My personal favourite smartphone app for stargazing. It's inexpensive and has a lovely user-friendly feel. You can set alerts for meteor showers, planetary conjunctions and passes of the International Space Station, too.

SkyView – This app is also easy to use and has some good features, including detailed information about each star, planet and constellation visible in the sky. The free version (SkyView Lite) is perfectly adequate for beginners.